A-Z
CELL AND
MOLECULAR BIOLOGY

A-Z CELL AND MOLECULAR BIOLOGY

Prof. Nirmal Chandra Pradhan

CENTRUM PRESS
NEW DELHI-110002 (INDIA)

CENTRUM PRESS
H.O.: 4360/4, Ansari Road, Daryaganj,
New Delhi-110 002 (India)
Ph.: 23278000, 23261597
B.O.: No. 1015, Ist Main Road, BSK IIIrd Stage
IIIrd Phase, IIIrd Block,
Bangalore - 560 085 (India)
Tel.: 080-41723429
Visit us at: www.centrumpress.com

A-Z Cell and Molecular Biology

First Edition, 2009
ISBN 978-93-80106-28-1

PRINTED IN INDIA

Printed at Salasar Imaging Systems, Delhi-110035 (India)

Contents

Preface

Molecular biology is the study of biology at a molecular level. The field overlaps with other areas of biology and chemistry, particularly genetics and biochemistry. Molecular biology chiefly concerns itself with understanding the interactions between the various systems of a cell, including the interactions between DNA, RNA and protein biosynthesis and learning how these interactions are regulated. One of the most basic techniques of molecular biology to study protein function is expression cloning.

In this technique, DNA coding for a protein of interest is cloned into a plasmid. This plasmid may have special promoter elements to drive production of the protein of interest, and may also have antibiotic resistance markers to help follow the plasmid.This plasmid can be inserted into either bacterial or animal cells.

Introducing DNA into bacterial cells can be done by transformation (via uptake of naked DNA), conjugation or by transduction. Introducing DNA into eukaryotic cells, such as animal cells, by physical or chemical means is called transfection. Several different transfection techniques are available, such as calcium phosphate transfection, electroporation, microinjection and liposome transfection. DNA can also be introduced into eukaryotic cells using viruses or bacteria as carriers, the latter is sometimes called bactofection and in particular uses Agrobacterium tumefaciens. Large quantities of a protein can then be extracted from the bacterial or eukaryotic cell.

Author

Preface

Molecular biology is the study of biology at a molecular level. The field overlaps with other areas of biology and chemistry, particularly genetics and biochemistry. Molecular biology chiefly concerns itself with understanding the interactions between the various systems of a cell, including the interactions between DNA, RNA and protein biosynthesis and learning how these interactions are regulated. [illegible]

[illegible]

Chapter 1

Genetic Molecule

The discovery that DNA is the prime genetic molecule, carrying all the hereditary information within chromosomes, immediately focused attention on its structure. It was hoped that knowledge of the structure would reveal how DNA carries the genetic messages that are replicated when chromosomes divide to produce two identical copies of themselves.

During the late 1940s and early 1950s, several research groups in the United States and in Europe engaged in serious efforts—both cooperative and rival—to understand how the atoms of DNA are linked together by covalent bonds and how the resulting molecules are arranged in three-dimensional space. Not surprisingly, there initially were fears that DNA might have very complicated and perhaps bizarre structures that differed radically from one gene to another. Great relief, if not general elation, was thus expressed when the fundamental DNA structure was found to be the double helix.

It told us that all genes have roughly the same three-dimensional form and that the differences between two genes reside in the order and number of their four nucleotide building blocks along the complementary strands. Now, some 50 years after the discovery of the double helix, this simple description of the genetic material remains true and has not had to be appreciably altered to accommodate new findings.

Nevertheless, we have come to realise that the structure of DNA is not quite as uniform as was first thought. For example, the chromosome of some small viruses have single-stranded, not double-stranded, molecules. Moreover, the precise orientation of the base pairs varies slightly from base pair to base pair in a manner that is influenced by the local DNA sequence. Some DNA sequences even permit the double helix to twist in the left-handed sense, as opposed to the right-handed sense originally formulated for DNA's general structure. And while some DNA molecules are linear, others are circular.

Still additional complexity comes from the supercoiling (further twisting) of the double helix, often around cores of DNA-binding proteins. Likewise, we now realise that RNA, which at first glance appears to be very similar to DNA, has its own distinctive structural features. It is principally found as a single-stranded molecule. Yet by means of intra-strand base pairing, RNA exhibits extensive double-helical character and is capable of folding into a wealth of diverse tertiary structures. These structures are full of surprises, such as non-classical base pairs, base-backbone interactions, and knot-like configurations. Most remarkable of all, and of profound evolutionary significance, some RNA molecules are enzymes that carry out reactions that are at the core of information transfer from nucleic acid to protein. Clearly, the structures of DNA and RNA are richer and more intricate than was at first appreciated. Indeed, there is no one generic structure for DNA and RNA.

DNA IS COMPOSED OF POLYNUCLEOTIDE CHAINS

The most important feature of DNA is that it is usually composed of two polynucleotide chains twisted around each other in the form of a double helix. The upper part of the figure presents the structure of the double helix shown in a schematic form. Note that if inverted 180°, the double helix looks superficially the same, due to the complementary nature of the two DNA strands. The space-filling model of the double helix, in the lower part of the figure, shows the components of the DNA molecule and their relative positions in the helical structure. The backbone of each strand of the helix is composed of alternating sugar and phosphate residues; the bases project inward but are accessible through the major and minor grooves.

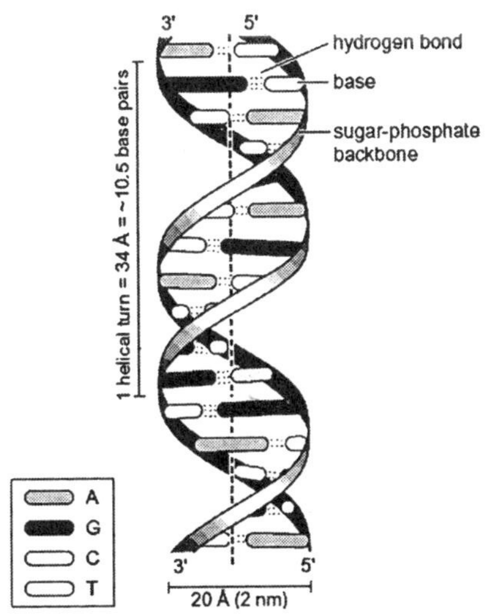

Fig. The Helical Structure of DNA

Let us begin by considering the nature of the nucleotide, the fundamental building block of DNA. The nucleotide consists of a phosphate joined to a sugar, known as 2′-deoxyribose, to which a base is attached. The sugar is called 2′-deoxyribose because there is no hydroxyl at position 2′ (just two hydrogens).

Note that the positions on the ribose are designated with primes to distinguish them from positions on the bases. We can think of how the base is joined to 2′-deoxyribose by imagining the removal of a molecule of water between the hydroxyl on the 1′ carbon of the sugar and the base to form a glycosidic bond.

The sugar and base alone are called a nucleoside. Likewise, we can imagine linking the phosphate to 2′-deoxyribose by removing a water molecule from between the phosphate and the hydroxyl on the 5′ carbon to make a 5′ phosphomonoester. Adding a phosphate (or more than one phosphate) to a nucleoside creates a nucleo*t*ide. Thus, by making a glycosidic bond between the base and the sugar, and by making a phosphoester bond between the sugar and the phosphoric acid, we have created a nucleotide. Nucleotides are, in turn, joined to each other in polynucleotide chains through the 3′ hydroxyl of 2′-deoxyribose of one nucleotide and the phosphate attached to the 5′ hydroxyl of another nucleotide.

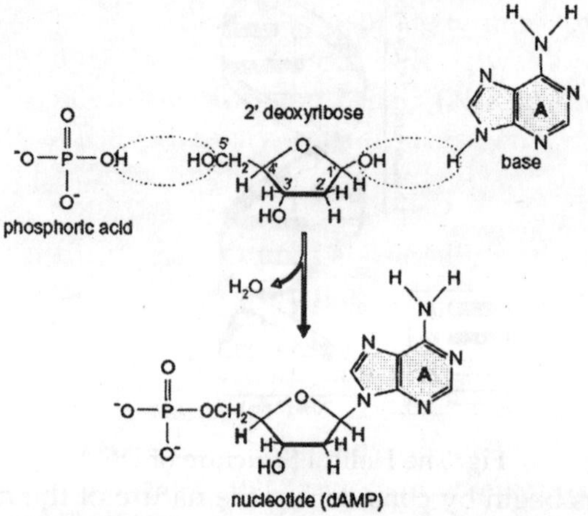

Fig. Formation of Nucleotide by Removal of Water

This is a phosphodiester linkage in which the phosphoryl group between the two nucleotides has one sugar esterified to it through a 3′ hydroxyl and a second sugar esterified to it through a 5′ hydroxyl. Phosphodiester linkages create the repeating, sugar-phosphate backbone of the polynucleotide chain, which is a regular feature of DNA. In contrast, the order of the bases along the polynucleotide chain is irregular. This irregularity as well as the long length is, as we shall see, the basis for the enormous information content of DNA. The phosphodiester linkages impart an inherent polarity to the DNA chain.

	Base Adenine	Nucleoside Adenosine	Nucleotide Adenosine 5'-phosphate	Deoxynucleotide Deoxyadenosine 5' phosphate
*Structure**	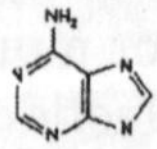	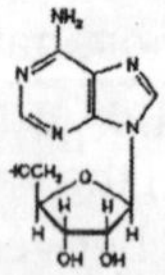	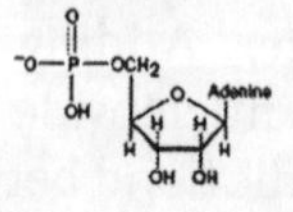	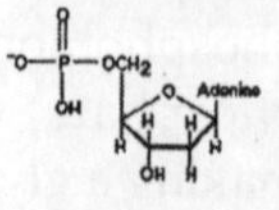
M.W.	135.1	267.2	347.2	331.2

This polarity is defined by the asymmetry of the nucleotides and the way they are joined. DNA chains have a free 5′ phosphate or 5′ hydroxyl at one end and a free 3′ phosphate or 3′ hydroxyl at the other end. The convention is to write DNA sequences from the 5′ end (on the left) to the 3′ end, generally with a 5′ phosphate and a 3′ hydroxyl.

EACH BASE HAS ITS PREFERRED TAUTOMERIC FORM

The bases in DNA fall into two classes, purines and pyrimidines. The purines are adenine and guanine, and the pyrimidines are cytosine and thymine. The purines are derived from the double-ringed structure. Adenine and guanine share this essential structure but with different groups attached. Likewise, cytosine and thymine are variations on the single-ringed structure shown in Figure. The figure also shows the numbering of the positions in the purine and pyrimidine rings. The bases are attached to the deoxyribose by glycosidic linkages at N1 of the pyrimidines or at N9 of the purines.

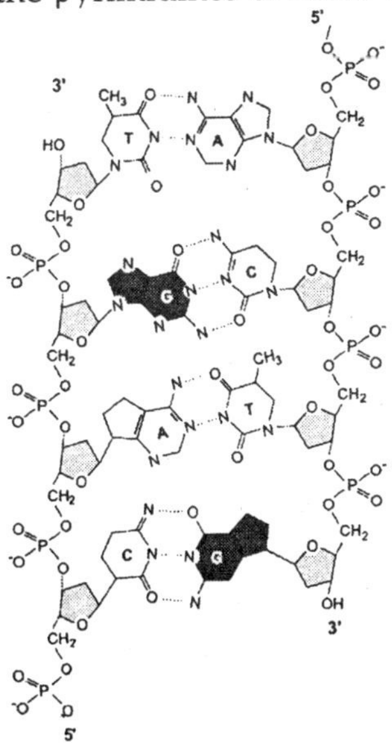

Fig. Detailed Structure of Polynucleotide Polymer

Each of the bases exists in two alternative tautomeric

states, which are in equilibrium with each other. The equilibrium lies far to the side of the conventional structures shown in Figure, which are the predominant states and the ones important for base pairing.

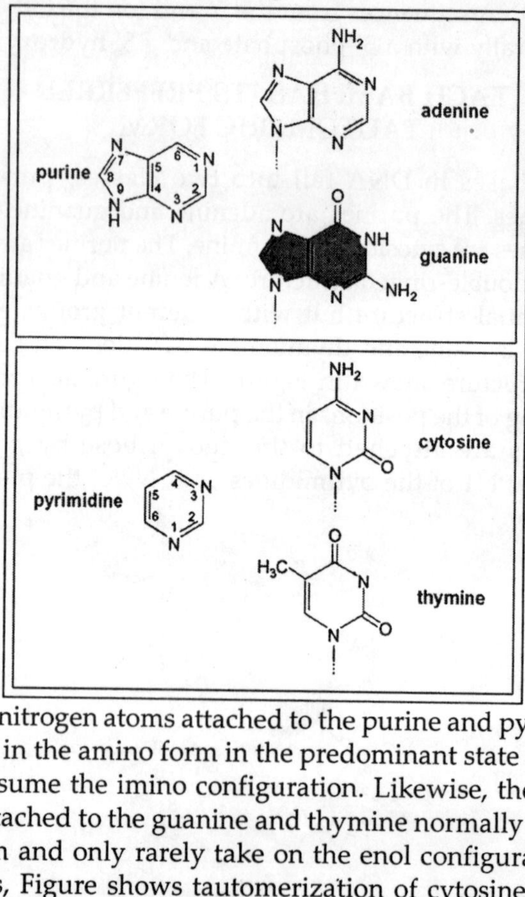

The nitrogen atoms attached to the purine and pyrimidine rings are in the amino form in the predominant state and only rarely assume the imino configuration. Likewise, the oxygen atoms attached to the guanine and thymine normally have the keto form and only rarely take on the enol configuration. As examples, Figure shows tautomerization of cytosine into the imino form (a) and guanine into the enol form (b). As we shall see, the capacity to form an alternative tautomer is a frequent source of errors during DNA synthesis.

STRANDS OF THE DOUBLE HELIX

The double helix is composed of two polynucleotide chains that are held together by weak, non-covalent bonds between pairs of bases, as shown in Figure. Adenine on one chain is always paired with thymine on the other chain and, likewise, guanine is always paired with cytosine. The two strands have the same helical geometry but base pairing holds them together with the opposite polarity.

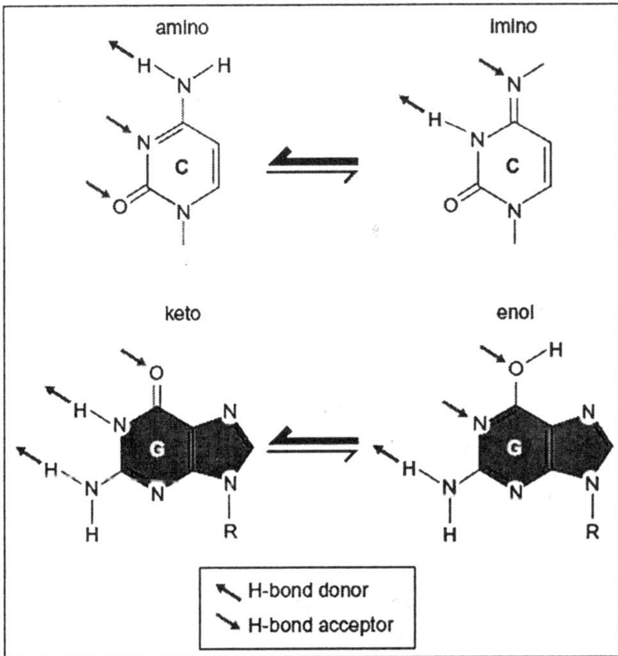

Fig. Base Tautomers

That is, the base at the 5′ end of one strand is paired with the base at the 3′ end of the other strand. The strands are said to have an anti-parallel orientation. This anti-parallel orientation is a stereochemical consequence of the way that adenine and thymine and guanine and cytosine pair with each together.

CHAINS OF THE DOUBLE HELIX

The pairing between adenine and thymine and between guanine and cytosine results in a complementary relationship between the sequence of bases on the two intertwined chains and gives DNA its self-encoding character. For example, if we have the sequence 5′-ATGTC-3′ on one chain, the opposite chain must have the complementary sequence 3′-TACAG-5′. The strictness of the rules for this "Watson-Crick" pairing derives from the complementarity both of shape and of hydrogen bonding properties between adenine and thymine and between guanine and cytosine. Adenine and thymine match up so that a hydrogen bond can form between the exocyclic amino group at on adenine and the carbonyl at in thymine; and likewise, a hydrogen bond can form between of adenine and of thymine.

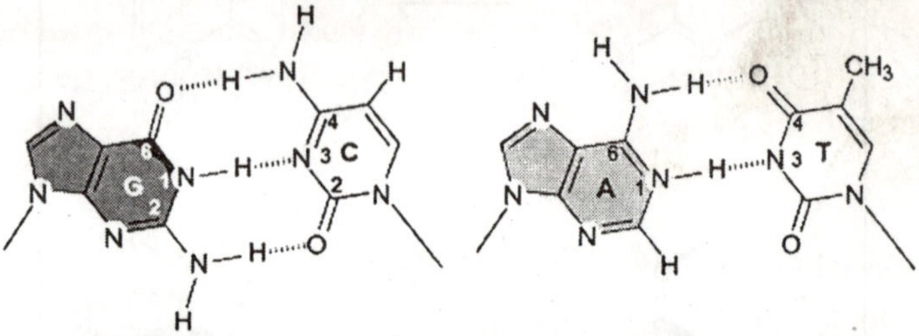

Fig. The Figure shows Hydrogen Bonding between the Bases

A corresponding arrangement can be drawn between a guanine and a cytosine, so that there is both hydrogen bonding and shape complementarity in this base pair as well. A G:C base pair has three hydrogen bonds, because the exocyclic NH_2 at on guanine lies opposite to, and can hydrogen bond with, a carbonyl at on cytosine.

Likewise, a hydrogen bond can form between of guanine and of cytosine and between the carbonyl at of guanine and the exocyclic NH_2 at C_4 of cytosine.

Watson-Crick base pairing requires that the bases are in their preferred tautomeric states. An important feature of the double helix is that the two base pairs have exactly the same geometry; having an A:T base pair or a G:C base pair between the two sugars does not perturb the arrangement of the sugars. Neither does T:A or C:G. In other words, there is an approximately twofold axis of symmetry that relates the two sugars and all four base pairs can be accommodated within the same arrangement without any distortion of the overall structure of the DNA.

HYDROGEN BONDING

The hydrogen bonds between complementary bases are a fundamental feature of the double helix, contributing to the thermodynamic stability of the helix and providing the information content and specificity of base pairing. Hydrogen bonding might not at first glance appear to contribute importantly to the stability of DNA for the following reason. An organic molecule in aqueous solution has all of its hydrogen bonding properties satisfied by water molecules that come on and off very rapidly.

As a result, for every hydrogen bond that is made when a base pair forms, a hydrogen bond with water is broken that was there before the base pair formed. Thus, the net energetic contribution of hydrogen bonds to the stability of the double helix would appear to be modest. However, when polynucleotide strands are separate, water molecules are lined up on the bases. When strands come together in the double helix, the water molecules are displaced from the bases.

This creates disorder and increases entropy, thereby stabilizing the double helix. Hydrogen bonds are not the only force that stabilizes the double helix. A second important contribution comes from stacking interactions between the bases. The bases are flat, relatively waterinsoluble molecules, and they tend to stack above each other roughly perpendicular to the direction of the helical axis. Electron cloud interactions (ð–ð) between bases in the helical stacks contribute significantly to the stability of the double helix. Hydrogen bonding is also important for the specificity of base pairing. Suppose we tried to pair an adenine with a cytosine.

Then we would have a hydrogen bond acceptor of adenine lying opposite a hydrogen bond acceptor of cytosine with no room to put a water molecule in between to satisfy the two acceptors. Likewise, two hydrogen bond donors, the NH_2 groups at of adenine and of cytosine, would lie opposite each other. Thus, an A:C base pair would be unstable because water would have to be stripped off the donor and acceptor groups without restoring the hydrogen bond formed within the base pair.

BASES CAN FLIP OUT FROM THE DOUBLE HELIX

As we have seen, the energetics of the double helix favour the pairing of each base on one polynucleotide strand with the complementary base on the other strand. Certain enzymes that methylate bases or remove damaged bases do so with the base in an extra helical configuration in which it is flipped out from the double helix, enabling the base to sit in the catalytic cavity of the enzyme.

Furthermore, enzymes involved in homologous recombination and DNA repair are believed to scan DNA for homology or lesions by flipping out one base after another. This is not energetically expensive because only one base is flipped out at a time. Clearly, DNA is more flexible than might be assumed at first glance.

DNA IS USUALLY A RIGHT-HANDED DOUBLE HELIX

Applying the handedness rule from physics, we can see that each of the polynucleotide chains in the double helix is right-handed. In your mind's eye, hold your right hand up to the DNA molecule in Figure with your thumb pointing up and along the long axis of the helix and your fingers following the grooves in the helix. Trace along one strand of the helix in the direction in which your thumb is pointing. Notice that you go around the helix in the same direction as your fingers are pointing.

This does not work if you use your left hand. Try it! A consequence of the helical nature of DNA is its periodicity. Each base pair is displaced (twisted) from the previous one by about 36°. Thus, in the X-ray crystal structure of DNA it takes a stack of about 10 base pairs to go completely around the helix (360°). That is, the helical periodicity is generally 10 base pairs per turn of the helix.

THE DOUBLE HELIX HAS MINOR AND MAJOR GROOVES

As a result of the double-helical structure of the two chains, the DNA molecule is a long extended polymer with two grooves that are not equal in size to each other. Why are there a minor groove and a major groove? It is a simple consequence of the geometry of the base pair. The angle at which the two sugars protrude from the base pairs (that is, the angle between the glycosidic bonds) is about 120° (for the narrow angle or 240° for the wide angle).

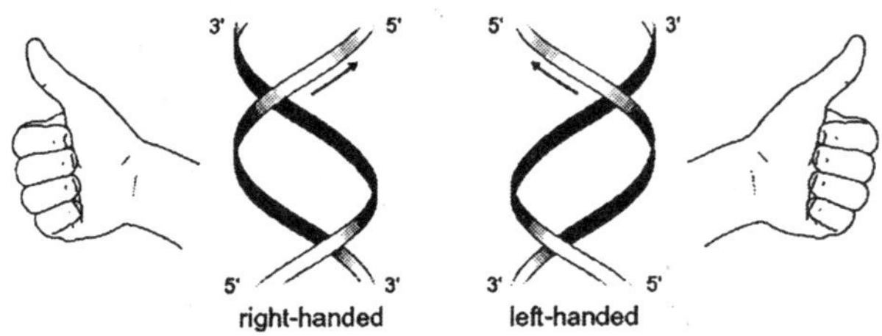

Fig. Left- and Right-Handed Helices

As a result, as more and more base pairs stack on top of each other, the narrow angle between the sugars on one edge of the base pairs generates a minor groove and the large angle on the other edge generates a major groove. (If the sugars pointed away from each other in a straight line, that is, at an angle of 180°, then two grooves would be of equal dimensions and there would be no minor and major grooves.)

THE MAJOR GROOVE IS RICH IN CHEMICAL INFORMATION

The edges of each base pair are exposed in the major and minor grooves, creating a pattern of hydrogen bond donors and acceptors and of van der Waals surfaces that identifies the base pair. The edge of an A:T base pair displays the following chemical groups in the following order in the major groove: a hydrogen bond acceptor (the of adenine), a hydrogen bond donor (the exocyclic amino group on of adenine), a hydrogen bond acceptor (the carbonyl group on of thymine) and a bulky hydrophobic surface (the methyl group on of thymine).

Similarly, the edge of a G:C base pair displays the following groups in the major groove: a hydrogen bond acceptor (at of guanine), a hydrogen bond acceptor (the carbonyl on of guanine), a hydrogen bond donor (the exocyclic amino group on of cytosine), a small non-polar hydrogen (the hydrogen at of cytosine). Thus, there are characteristic patterns of hydrogen bonding and of overall shape that are exposed in the major groove that distinguish an A:T base pair from a G:C base pair, and, for that matter, A:T from T:A, and G:C from C:G. We can think of these features as a code in which A represents a hydrogen bond acceptor, D a hydrogen bond donor, M a methyl group, and H a nonpolar hydrogen. In such a code, A D A M in the major groove signifies an A:T base pair, and A A D H stands for a G:C base pair.

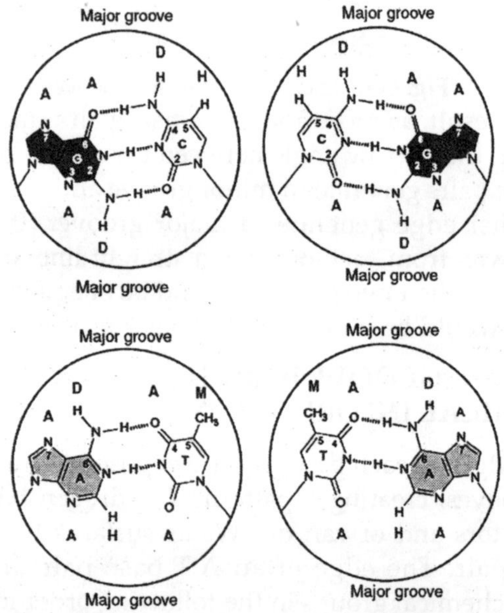

Fig. Chemical Groups Exposed in the Major and Minor Grooves from the Edges of the Base Pairs. The Letters Identify as Hydrogen Bond Acceptors (A), Hydrogen Bond Donors (D), Nonpolar Hydrogens (H), and Methyl groups (M).

Likewise, **M A D A** stands for a T:A base pair and **H D A A** is characteristic of a C:G base pair. In all cases, this code of chemical groups in the major groove specifies the identity of the base pair. These patterns are important because they allow proteins to unambiguously recognize DNA sequences without having to open and thereby disrupt the double helix. Indeed, as we shall see, a principal decoding mechanism relies upon the ability of amino acid side chains to protrude into the major groove and to recognize and bind to specific DNA sequences.

The minor groove is not as rich in chemical information and what information is available is less useful for distinguishing between base pairs. The small size of the minor groove is less able to accommodate amino acid side chains. Also, A:T and T:A base pairs and G:C and C:G pairs look similar to one another in the minor groove.

An A:T base pair has a hydrogen bond acceptor (at of adenine), a nonpolar hydrogen (at of adenine) and a hydrogen bond acceptor (the carbonyl on of thymine). Thus, its code is A H A. But this code is the same if read in the opposite direction, and hence an A:T base pair does not look very different from a T:A base pair from the point of view of the hydrogenbonding properties of a protein poking its side chains into the minor groove.

Likewise, a G:C base pair exhibits a hydrogen bond acceptor (at N3 of guanine), a hydrogen bond donor (the exocyclic amino group on guanine), and a hydrogen bond acceptor (the carbonyl on of cytosine), representing the code A D A. Thus, from the point of view of hydrogen bonding, C:G and G:C base pairs do not look very different from each other either. The minor groove does look different when comparing an A:T base pair with a G:C base pair, but G:C and C:G, or A:T and T:A, cannot be easily distinguished.

THE DOUBLE HELIX EXISTS IN MULTIPLE CONFORMATIONS

Early X-ray diffraction studies of DNA, which were carried out using concentrated solutions of DNA that had been drawn out into thin fibers, revealed two kinds of structures, the B and the A forms of DNA. The B form, which is observed at high humidity, most closely corresponds to the average structure of DNA under physiological conditions. It has 10 base pairs per turn, and a wide major groove and a narrow minor groove.

The A form, which is observed under conditions of low humidity, has 11 base pairs per turn. Its major groove is narrower and much deeper than that of the B form, and its minor groove is broader and shallower. The vast majority of the DNA in the cell is in the B form, but DNA does adopt the A structure in certain DNA-protein complexes. Also, as we shall see, the A form is similar to the structure that RNA adopts when double helical.

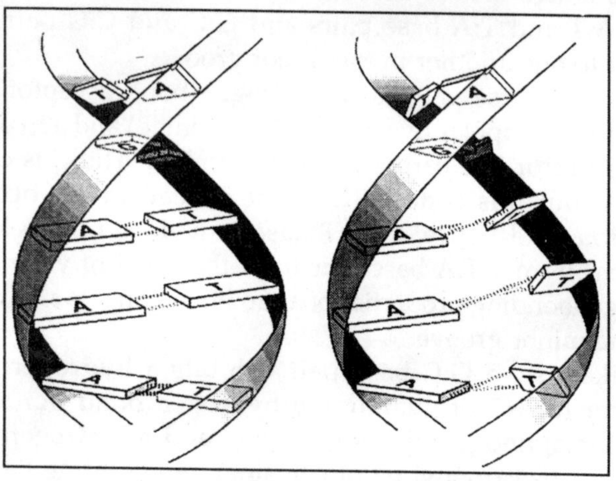

Fig. The Propeller Twist between the Purine and Pyrimidine Base Pairs of a Right-Handed Helix

The B form of DNA represents an ideal structure that deviates in two respects from the DNA in cells. First, DNA in solution, as we have seen, is somewhat more twisted on average than the B form, having on average 10.5 base pairs per turn of the helix. Second, the B form is an average structure whereas real DNA is not perfectly regular. Rather, it exhibits variations in its precise structure from base pair to base pair. This was revealed by comparison of the crystal structures of individual DNAs of different sequences. For example, the two members of each base pair do not always lie exactly in the same plane.

Rather, they can display a "propeller twist" arrangement in which the two flat bases counter rotate relative to each other along the long axis of the base pair, giving the base pair a propeller-like character. Moreover, the precise rotation per base pair is not a constant. As a result, the width of the major and minor grooves varies locally. Thus, DNA molecules are never perfectly regular double helices. Instead, their exact conformation depends on which base pair (A:T, T:A, G:C, or C:G) is present at each position along the double helix and on the identity of neighboring base pairs. Still, the B form is for many purposes a good first approximation of the structure of DNA in cells.

DNA CAN SOMETIMES FORM A LEFT-HANDED HELIX

DNA containing alternative purine and pyrimidine residues can fold into left-handed as well as right-handed helices. To understand how DNA can form a left-handed helix, we need to consider the glycosidic bond that connects the base to the 1′ position of 2′-deoxyribose. This bond can be in one of two conformations called *syn* and *anti*.

In right-handed DNA, the glycosidic bond is always in the *anti* conformation. In the left-handed helix, the fundamental repeating unit usually is a purine-pyrimidine dinucleotide, with the glycosidic bond in the *anti* conformation at pyrimidine residues and in the *syn* conformation at purine residues. It is this *syn* conformation at the purine nucleotides that is responsible for the left-handedness of the helix.

The change to the *syn* position in the purine residues to alternating *anti–syn* conformations gives the backbone of left-handed DNA a zigzag look, which distinguishes it from right-handed forms. The rotation that effects the change from *anti* to *syn* also causes the ribose group to undergo a change in its pucker. In solution alternating purine–pyrimidine residues assume the left-handed conformation only in the presence of high concentrations of positively charged ions (e.g., Na′) that shield the negatively charged phosphate groups.

At lower salt concentrations, they form typical right-handed conformations. The physiological significance of Z DNA is uncertain and left-handed helices probably account at most for only a small of proportion of a cell's DNA.

DNA STRANDS CAN SEPARATE (DENATURE) AND REASSOCIATE

Because the two strands of the double helix are held together by relatively weak (non-covalent) forces, you might expect that the two strands could come apart easily. Indeed, the original structure for the double helix suggested that DNA replication would occur in just this manner. The complementary strands of double helix can also be made to come apart when a solution of DNA is heated above physiological temperatures (to near 100 °C) or under conditions of high pH, a process known as denaturation. However, this complete separation of DNA strands by denaturation is reversible.

When heated solutions of denatured DNA are slowly cooled, single strands often meet their complementary strands and reform regular double helices. The capacity to renature denatured DNA molecules permits artificial hybrid DNA molecules to be formed by slowly cooling mixtures of denatured DNA from two different sources. Likewise, hybrids can be formed between complementary strands of DNA and RNA.

The ability to form hybrids between two single-stranded nucleic acids (hybridization) is the basis for several indispensable techniques in molecular biology, such as Southern blot hybridization and DNA microarrays. Important insights into the properties of the double helix were obtained from classic experiments carried out in the 1950s in which the denaturation of DNA was studied under a variety of conditions. In these experiments DNA denaturation was monitored by measuring the absorbance of ultraviolet light passed through a solution of DNA. DNA maximally absorbs ultraviolet light at a wavelength of about 260 nm.

It is the bases that are principally responsible for this absorption. When the temperature of a solution of DNA is raised to near the boiling point of water, the optical density (absorbance) at 260 nm markedly increases. The explanation for this increase is that duplex DNA is hypochromic; it absorbs less ultraviolet light by about 40% than do individual DNA chains. The hypochromicity is due to base stacking, which diminishes the capacity of the bases in duplex DNA to absorb ultraviolet light. If we plot the optical density of DNA as a function of temperature, we observe that the increase in absorption occurs abruptly over a relatively narrow temperature range.

The midpoint of this transition is the melting point or Tm. Like ice, DNA melts: it undergoes a transition from a highly ordered double-helical structure to a much less ordered structure of individual strands.

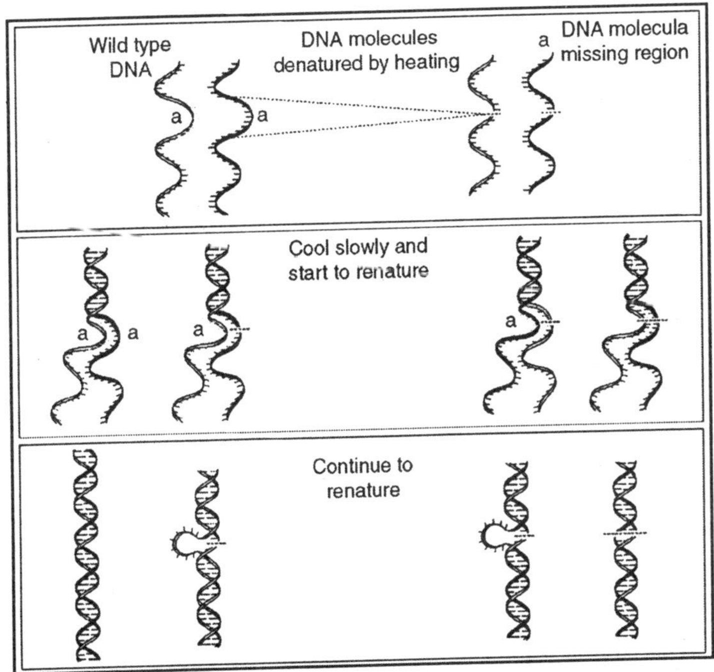

Fig. Reannealing and Hybridization

The sharpness of the increase in absorbance at the melting temperature tells us that the denaturation and renaturation of complementary DNA strands is a highly cooperative, zippering-like process. Renaturation, for example, probably occurs by means of a slow nucleation process in which a relatively small stretch of bases on one strand find and pair with their complement on the complementary strand. The remainder of the two strands then rapidly zipper-up from the nucleation site to reform an extended double helix.

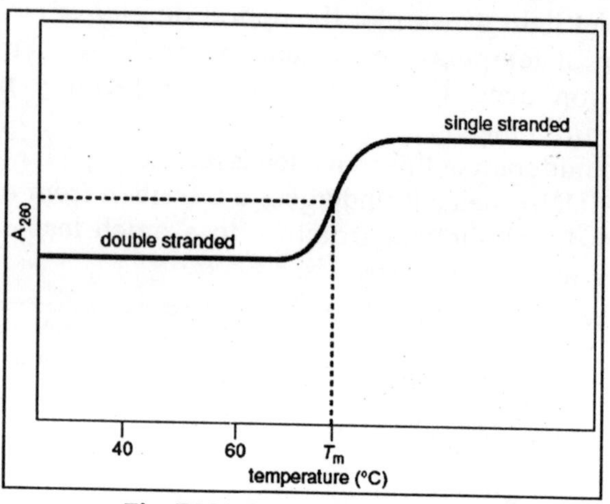

Fig. DNA Denaturation Curve

The melting temperature of DNA is a characteristic of each DNA that is largely determined by the G:C content of the DNA and the ionic strength of the solution. The higher the percent of G:C base pairs in the DNA (and hence the lower the content of A:T base pairs), the higher the melting point. Likewise, the higher the salt concentration of the solution the greater the temperature at which the DNA denatures.

How do we explain this behaviour? G:C base pairs contribute more to the stability of DNA than do A:T base pairs because of the greater number of hydrogen bonds for the former (three in a G:C base pair versus two for A:T) but also importantly because the stacking interactions of G:C base pairs with adjacent base pairs are more favorable than the corresponding interactions of A:T base pairs with their neighboring base pairs. The effect of ionic strength reflects another fundamental feature of the double helix.

The backbones of the two DNA strands contain phosphoryl groups, which carry a negative charge. These negative charges are close enough across the two strands that if not shielded they tend to cause the strands to repel each other, facilitating their separation. At high ionic strength, the negative charges are shielded by cations, thereby stabilizing the helix. Conversely, at low ionic strength the unshielded negative charges render the helix less stable.

CIRCLES CIRCULAR DNA MOLECULES

It was initially believed that all DNA molecules are linear and have two free ends. Indeed, the chromosomes of eukaryotic cells each contain a single (extremely long) DNA molecule. But now we know that some DNAs are circles. For example, the chromosome of the small monkey DNA virus SV40 is a circular, double-helical DNA molecule of about 5,000 base pairs. Also, most (but not all) bacterial chromosomes are circular; *E. coli* has a circular chromosome of about 5 million base pairs.

Additionally, many bacteria have small autonomously replicating genetic elements known as plasmids, which are generally circular DNA molecules. Interestingly, some DNA molecules are sometimes linear and sometimes circular. The most well-known example is that of the bacteriophage ', a DNA virus of *E. coli.*

The phage ' genome is a linear double-stranded molecule in the virion particle. However, when the ' genome is injected into an *E. coli* cell during infection, the DNA circularizes. This occurs by base-pairing between single-stranded regions that protrude from the ends of the DNA and that have complementary sequences ("sticky ends").

DNA TOPOLOGY

As DNA is a flexible structure, its exact molecular parameters are a function of both the surrounding ionic environment and the nature of the DNA-binding proteins with which it is complexed. Because their ends are free, linear DNA molecules can freely rotate to accommodate changes in the number of times the two chains of the double helix twist about each other.

But if the two ends are covalently linked to form a circular DNA molecule and if there are no interruptions in the sugar phosphate backbones of the two strands, then the absolute number of times the chains can twist about each other cannot change. Such a covalently closed, circular DNA is said to be topologically constrained. Even the linear DNA molecules of eukaryotic chromosomes are subject to topological constraints due to their entrainment in chromatin and interaction with other cellular components.

Despite these constraints, DNA participates in numerous dynamic processes in the cell. For example, the two strands of the double helix, which are twisted around each other, must rapidly separate in order for DNA to be duplicated and to be transcribed into RNA. Thus, understanding the topology of DNA and how the cell both accommodates and exploits topological constraints during DNA replication, transcription, and other chromosomal transactions is of fundamental importance in molecular biology.

LINKING NUMBER OF COVALENTLY CLOSED, CIRCULAR DNA

Let us consider the topological properties of covalently closed, circular DNA, which is referred to as cccDNA. Because there are no interruptions in either polynucleotide chain, the two strands of cccDNA cannot be separated from each other without the breaking of a covalent bond.

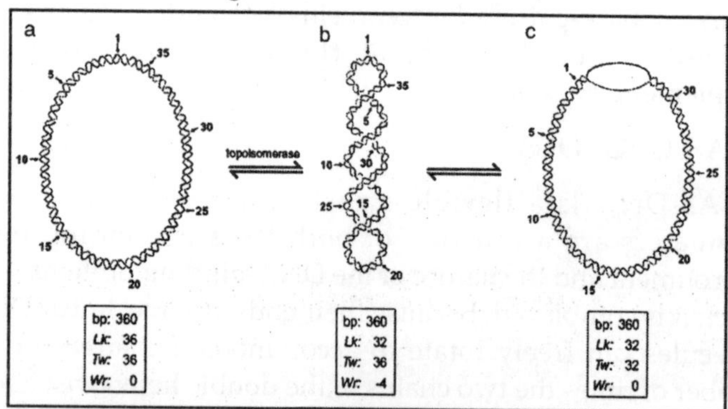

Fig. Topological States of Covalently Closed Circular (ccc) DNA

If we wished to separate the two circular strands without permanently breaking any bonds in the sugar phosphate backbones, we would have to pass one strand through the other strand repeatedly(we will encounter an enzyme that can perform just this feat!).

The number of times one strand would have to be passed through the other strand in order for the two strands to be entirely separated from each other is called the linking number. The linking number, which is always an integer, is an invariant topological property of cccDNA, no matter how much the DNA molecule is distorted.

Linking Number Is Composed of Twist and Writhe

The linking number is the sum of two geometric components called the twist and the writhe. Let us consider twist first. Twist is simply the number of helical turns of one strand about the other, that is, the number of times one strand completely wraps around the other strand. Consider a cccDNA that is lying flat on a plane.

In this flat conformation, the linking number is fully composed of twist. Indeed, the twist can be easily determined by counting the number of times the two strands cross each other. The helical crossovers (twist) in a right-handed helix are defined as positive such that the linking number of DNA will have a positive value. But cccDNA is generally not lying flat on a plane.

Rather, it is usually torsionally stressed such that the long axis of the double helix crosses over itself, often repeatedly, in three-dimensional space. This is called *writhe*. To visualize the distortions caused by torsional stress, think of the coiling of a telephone cord that has been overtwisted. Writhe can take two forms. One form is the interwound or plectonemic writhe, in which the long axis is twisted around itself.

The other form of writhe is a toroid or spiral in which the long axis is wound in a cylindrical manner, as often occurs when DNA wraps around protein. The writhing number (*Wr*) is the total number of interwound and/or spiral writhes in cccDNA. For example, the molecule shown in Figure has a writhe of 4 from 4 interwound writhes. Interwound writhe and spiral writhe are topologically equivalent to each other and are readily interconvertible geometric properties of cccDNA.

Also, twist and writhe are interconvertible. A molecule of cccDNA can readily undergo distortions that convert some of its twist to writhe or some of its writhe to twist without the breakage of any covalent bonds.

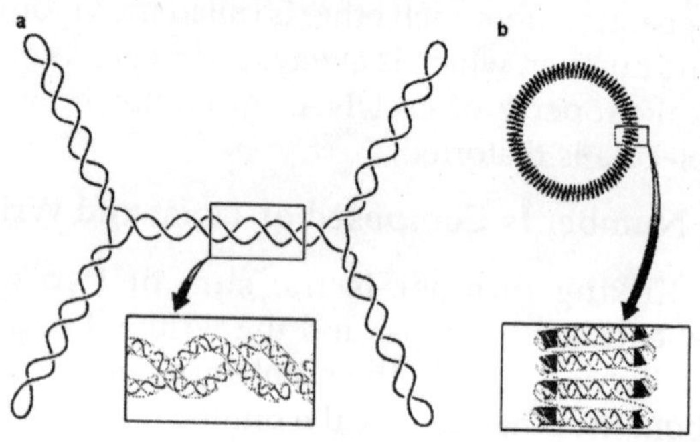

Fig. Two Forms of Writhe of Supercoiled DNA

The only constraint is that the sum of the twist number (*Tw*) and the writhing number (*Wr*) must remain equal to the linking number (*Lk*). This constraint is described by the equation: $Lk = Tw + Wr$.

Lk° Is the Linking Number of Fully Relaxed cccDNA

Consider cccDNA that is free of supercoiling (that is, it is said to be relaxed) and whose twist corresponds to that of the B form of DNA in solution under physiological conditions (about 10.5 base pairs per turn of the helix). The linking number (*Lk*) of such cccDNA under physiological conditions is assigned the symbol $Lk°$. $Lk°$ for such a molecule is the number of base pairs divided by 10.5.

For a cccDNA of 10,500 base pairs, $Lk = +1{,}000$. (The sign is positive because the twists of DNA are right-handed.) One way to see this is to imagine pulling one strand of the 10,500 base pair cccDNA out into a flat circle. If we did this, then the other strand would cross the flat circular strand 1,000 times.

How can we remove supercoils from cccDNA if it is not already relaxed? One procedure is to treat the DNA mildly with the enzyme DNase I, so as to break on average one phosphodiester bond (or a small number of bonds) in each DNA molecule. Once the DNA has been "nicked" in this manner, it is no longer topologically constrained and the strands can rotate freely, allowing writhe to dissipate.

If the nick is then repaired, the resulting cccDNA molecules will be relaxed and will have on average an *Lk* that is equal to Lk°. (Due to rotational fluctuation at the time the nick is repaired, some of

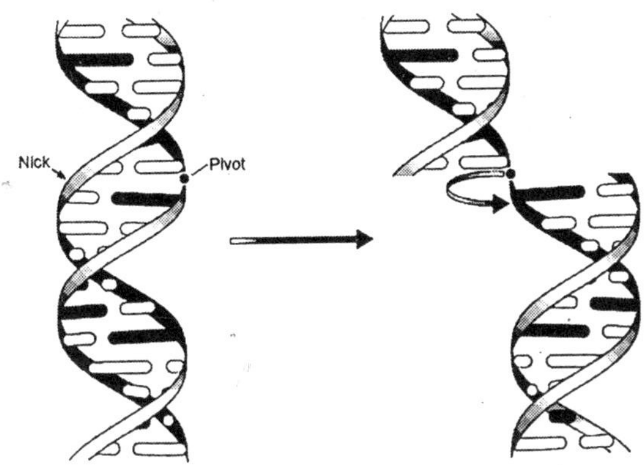

Fig. Relaxing DNA with DNase I

the resulting cccDNAs will have an *Lk* that is somewhat greater than Lk° and others will have an *Lk* that is somewhat lower. Thus, the relaxation procedure will generate a narrow spectrum of topoisomers whose average *Lk* is equal to Lk°).

DNA in Cells is Negatively Supercoiled

The extent of supercoiling is measured by the difference between *Lk* and Lk°, which is called the linking difference:

$$Lk = Lk - Lk^{\circ}.$$

If the "*Lk* of a cccDNA is significantly different from zero, then the DNA is torsionally strained and hence it is supercoiled. If $Lk < Lk^{\circ}$ and "$Lk < 0$, then the DNA is said to be "negatively supercoiled." Conversely, if $Lk < Lk^{\circ}$ and " $Lk < 0$, then the DNA is "positively supercoiled." For example, the molecule shown in Figure is negatively supercoiled and has a linking difference of -4 because its *Lk* is four less than that for the relaxed form of the molecule shown in Figure.

Because " *Lk* and Lk^{O} are dependent upon the length of the DNA molecule, it is more convenient to refer to a normalized measure of supercoiling. This is the superhelical density, which is assigned the symbol ó and is defined as:

$$ó = "Lk/Lk^{\circ}$$

DNA rings purified both from bacteria and eukaryotes are usually negatively supercoiled, having values of ó of about -0.06. The electron micrograph shown in Figure compares the structures of bacteriophage DNA in its relaxed form with its supercoiled form. What does this mean biologically? Negative supercoils can be thought of as a store of free energy that aids in processes that require strand separation, such as DNA replication and transcription. Because

$$Lk = Tw + Wr,$$

negative supercoils can be converted into untwisting of the double helix. Regions of negatively supercoiled DNA therefore have a tendency to partially unwind. Thus, strand separation can be accomplished more easily in negatively supercoiled DNA than in relaxed DNA. The only organisms that have been found to have positively supercoiled DNA are certain thermophiles, microorganisms that live under conditions of extreme high temperatures, such as in hot springs.

In this case, the positive supercoils can be thought of as a store of free energy that helps keep the DNA from denaturing at the elevated temperatures. In so far as positive supercoils can be converted into more twist (positively supercoiled DNA can be thought of as being overwound), strand separation requires more energy in thermophiles than in organisms whose DNA is negatively supercoiled.

Nucleosomes Introduce Negative Supercoiling in Eukaryotes

DNA in the nucleus of eukaryotic cells is packaged in small particles known as nucleosomes in which the double helix is wrapped almost two times around the outside circumference of a protein core. You will be able to recognize this wrapping as the toroid or spiral form of writhe. Importantly, it occurs in a lefthanded manner.. It turns out that writhe in the form of left-handed spirals is equivalent to negative supercoils. Thus, the packaging of DNA into nucleosomes introduces negative superhelical density.

Topoisomerases Can Relax Supercoiled DNA

As we have seen, the linking number is an invariant property of DNA that is topologically constrained. It can only be changed by introducing interruptions into the sugar-phosphate backbone. A remarkable class of enzymes known as topoisomerases are able to do just that by introducing transient nicks or breaks into the DNA. Topoisomerases are of two broad types.

Type II topoisomerases change the linking number in steps of two. They make transient double-stranded breaks in the DNA, through which they pass a region of uncut duplex DNA before resealing the break. Type II topoisomerases require energy from ATP hydrolysis for their action. Type I topoisomerases, in contrast, change the linking number of DNA in steps of one.

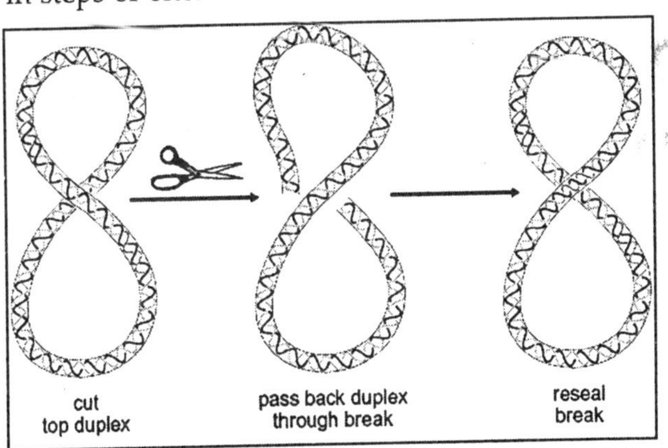

Fig. Schematic for Changing the Linking Number in DNA with Topoisomerase II

They make transient singlestranded breaks in the DNA, allowing one strand to pass through the break in the other before resealing the nick. Type I topoisomerases relax DNA by removing supercoils (dissipating writhe).

They can be compared to the protocol of introducing nicks into cccDNA with DNase and then repairing the nicks, which as we saw can be used to relax cccDNA, except that type I topoisomerases relax DNA in a controlled and concerted manner. In contrast to type II topoisomerases, type I topoisomerases do not require ATP. Both type I and type II topoisomerases work through an intermediate in which the enzyme is covalently attached to one end of the broken DNA.

Prokaryotes Have a Special Topoisomerase that Introduces Supercoils

Both prokaryotes and eukaryotes have type I and type II topoisomerases, which are capable of removing supercoils from DNA. In addition, however, prokaryotes have a special type II topoisomerase known as DNA gyrase that is able to introduce negative supercoils, rather than remove them. DNA gyrase is responsible for the negative supercoiling of chromosomes in prokaryotes, which facilitates unwinding of the DNA duplex during transcription and DNA replication.

DNA Topoisomers Can Be Separated by Electrophoresis

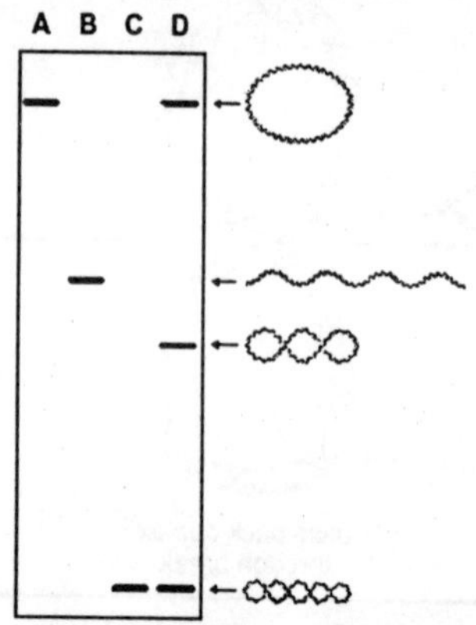

Fig. Schematic of Electrophoretic Separation of DNA Topoisomers

Covalently closed, circular DNA molecules of the same length but of different linking numbers are called DNA topoisomers. Even though topoisomers have the same molecular weight, they can be separated from each other by electrophoresis through a gel of agarose. The basis for this separation is that the greater the writhe the more compact the shape of a cccDNA. Once again, think of how supercoiling a telephone cord causes it to become more compact. The more compact the DNA, the more easily (up to a point) it is able to migrate through the gel matrix.

Thus, a fully relaxed cccDNA migrates more slowly than a highly supercoiled topoisomer of the same circular DNA. Figure shows a ladder of DNA topoisomers resolved by gel electrophoresis. Molecules in adjacent rungs of the ladder differ from each other by a linking number difference of just one. Obviously, electrophoretic mobility is highly sensitive to the topological state of DNA.

Proving That DNA Has a Helical Periodicity of about 10.5 Base

The observation that DNA topoisomers can be separated from each other electrophoretically is the basis for a simple experiment that proves that DNA has a helical periodicity of about 10.5 base pairs per turn in solution. Consider three cccDNAs of sizes 3990, 3995, and 4011 base pairs that were relaxed to completion by treatment with topoisomerase I. When subjected to electrophoresis through agarose, the 3990- and 4011-base-pair DNAs exhibit essentially identical mobilities.

Due to thermal fluctuation, topoisomerase treatment actually generates a narrow spectrum of topoisomers, but for simplicity let us consider the mobility of only the most abundant topoisomer (that corresponding to the cccDNA in its most relaxed state). The mobilities of the most abundant topoisomers for the 3990- and 4011-base-pair DNAs are indistinguishable because the 21-base-pair difference between them is negligible compared to the sizes of the rings. The most abundant topoisomer for the 3995-base-pair ring, however, is found to migrate slightly more rapidly than the other two rings even though it is only 5 base pairs larger than the 3990-base-pair ring.

How are we to explain this anomaly? The 3990- and 4011-base-pair rings in their most relaxed states are expected to have linking numbers equal to *Lk*°, that is, 380 in the case of the 3990-base-pair ring (dividing the size by 10.5 base pairs) and 382 in the case of the 4011-base-pair ring. Because *Lk* is equal to *Lk*°, the linking difference ('*Lk* = *Lk* - *Lk*°) in both cases is zero and there is no writhe. But because the linking number must be an integer, the most relaxed state for the 3995-base-pair ring would be either of two topoisomers having linking numbers of 380 or 381. However, *Lk*° for the 3995-base-pair ring is 380.5.

Thus, even in its most relaxed state, a covalently closed circle of 3995 base pairs would necessarily have about half a unit of writhe (its linking difference would be 0.5), and hence it would migrate more rapidly than the 3990- and 4011-base-pair circles. In other words, to explain how rings that differ in length by 21 base pairs (two turns of the helix) have the same mobility whereas a ring that differs in length by only 5 base pairs (about half a helical turn) exhibits a different mobility, we must conclude that DNA in solution has a helical periodicity of about 10.5 base pairs per turn.

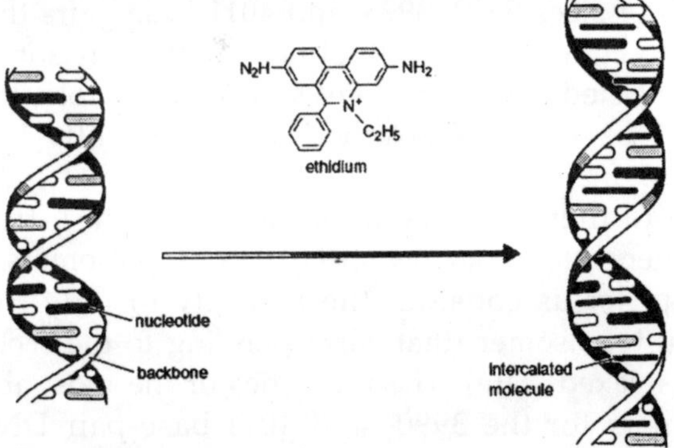

Fig. Intercalation of Ethidium Bromide into DNA

Ethidium Ions Cause DNA to Unwind

Ethidium is a large, flat, multi-ringed cation. Its planar shape enables ethidium to slip (intercalate) between the stacked base pairs of DNA. Because it fluoresces when exposed to ultraviolet light, and because its fluorescence increases dramatically after intercalation, ethidium is used as a stain to visualize DNA. When an ethidium ion intercalates between two base pairs, it causes the DNA to unwind by 26°, reducing the normal rotation per base pair from ~36° to ~10°. In other words, ethidium decreases the twist of DNA.

Imagine the extreme case of a DNA molecule that has an ethidium ion between every base pair. Instead of 10 base pairs per turn it would have 36! When ethidium binds to linear DNA or to a nicked circle, it simply causes the helical pitch to increase. But consider what happens when ethidium binds to covalently closed, circular DNA. The linking number of the cccDNA does not change (no covalent bonds are broken and resealed), but the twist decreases by 26° for each molecule of ethidium that has bound to the DNA.

Because $Lk = Tw + Wr$, this decrease in *Tw* must be compensated for by a corresponding increase in *Wr*. If the circular DNA is initially negatively supercoiled (as is normally the case for circular DNAs isolated from cells), then the addition of ethidium will increase *Wr*. In other words, the addition of ethidium will relax the DNA. If enough ethidium is added, the negative supercoiling will be brought to zero, and if even more ethidium is added, *Wr* will increase above zero and the DNA will become positively supercoiled.

Because the binding of ethidium increases *Wr*, its presence greatly affects the migration of cccDNA during gel electrophoresis. In the presence of non-saturating amounts of ethidium, negatively supercoiled circular DNAs are more relaxed and migrate more slowly, whereas relaxed cccDNAs become positively supercoiled and migrate more rapidly.

Chapter 2

DNA Structure

DNA stands for deoxyribonucleic acid. DNA is pretty unusual in that it is about the only common molecule capable of directing its own synthesis. The processes of mitosis and meiosis were discovered in the 1870s and 1890s. It was observed that, as cells divided, chromosomes moved around in a cell, and people began to wonder what their function was.

It was determined that chromosomes were made of protein and DNA, about which people knew almost nothing. People began to suspect that chromosomes had something to do with genetics, but couldn't explain what/how. When enough evidence was accumulated to confirm that chromosomes did, indeed, have something to do with genetics, most people thought that in some way the protein in the chromosomes served as the genetic material.

People knew that DNA was also in the chromosomes, but because its structure was unknown and people didn't know much about it, few people thought it was the genetic material. Frederick Griffith performed an experiment using pneumonia bacteria and mice. This was one of the first experiments that hinted that DNA was the genetic code material. He used two strains of *Streptococcus pneumoniae*: a "smooth" strain which has a polysaccharide coating around it that makes it look smooth when viewed with a microscope, and a "rough" strain which doesn't have the coating, thus looks rough under the microscope.

When he injected live S strain into mice, the mice contracted pneumonia and died. When he injected live R strain, a strain which typically does not cause illness, into mice, as predicted they did not get sick, but lived. Thinking that perhaps the polysaccharide coating on the bacteria somehow caused the illness and knowing that polysaccharides are not affected by heat, Griffith then used heat to kill some of the S strain bacteria and injected those dead bacteria into mice.

This failed to infect/kill the mice, indicating that the polysaccharide coating was not what caused the disease, but rather, something within the living cell. Since Griffith had used heat to kill the bacteria and heat denatures protein, he next hypothesized that perhaps some protein within the living cells, that was denatured by the heat, caused the disease. He then injected another group of mice with a mixture of heat-killed S and live R, and the mice died! When he did a necropsy on the dead mice, he isolated live S strain bacteria from the corpses.

Griffith concluded that the live R strain bacteria must have absorbed genetic material from the dead S strain bacteria, and since heat denatures protein, the protein in the bacterial chromosomes was not the genetic material. This evidence pointed to DNA as being the genetic material. Transformation is the process whereby one strain of a bacterium absorbs genetic material from another strain of bacteria and "turns into" the type of bacterium whose genetic material it absorbed. Because DNA was so poorly understood, scientists remained skeptical up through the 1940s.

Alfred Hershey and Martha Chase did an experiment which is so significant, it has been nicknamed the "Hershey-Chase Experiment". At that time, people knew that viruses were composed of DNA (or RNA) inside a protein coat/shell called a capsid. It was also known that viruses replicate by taking over the host cell's metabolic functions to make more virus. We are used to thinking and talking about viruses which invade our bodies and make us sick, but there are other, different kinds of viruses that infect other kinds of animals, still other viruses which infect plants, and even some viruses that infect bacteria.

A virus which infects a bacterium is called a bacteriophage because the host bacterium cell is killed as the new virus particles leave the bacterial cell. In order to do all this, the virus must inject whatever is the viral genetic code into the host cell.

Thus, people realized that the viral genetic code material had to be either its DNA or its protein capsid. Hershey and Chase sought an answer to the question, "Is it the viral DNA or viral protein coat (capsid) that is the viral genetic code material which gets injected into a host bacterium cell? To try to answer this question, Hershey and Chase performed an experiment using a bacterium named *Escherichia coli*, or *E. coli* for short (named after a scientist whose last name was Escher) and a virus called that is a bacteriophage that infects *E. coli*. Isolated, like other viruses, is just a crystal of DNA and protein, so it must live inside *E. coli* in order to make more virus like itself.

When the new viruses are ready to leave the host *E. coli* cell (and go infect others), they burst the *E. coli* cell open, killing it (hence the name "bacteriophage"). The results that Hershey and Chase obtained indicated that the viral DNA, not the protein, is its genetic code material.

Hershey and Chase used radioactive chemicals to distinguish between ("label") the protein capsid and the DNA in virus so they could tell which of those molecules entered the *E. coli* cells. Since some amino acids contain sulfur in their side chains, if is grown in *E. coli* with a source of radioactive sulfur, the sulfur will be incorporated into the protein coat making it radioactive. Since DNA has lots of phosphorus in its phosphate ($-PO_4$) groups, if is grown in *E. coli* with a source of radioactive phosphorus, the phosphorus will be incorporated into the viral DNA, making that radioactive.

Hershey and Chase grew two batches of and *E. coli*: one with radioactive sulfur and one with radioactive phosphorus to get batches of "labeled" with either radioactive S or radioactive P. Then, these radioactive were placed in separate, new batches of *E. coli*, but were left there only 10 minutes. This was to give the time to inject their genetic material into the bacteria, but not reproduce.

In the next step, still in separate batches, the mixtures were agitated in a kitchen blender to knock loose any viral parts not inside the *E. coli* but perhaps stuck on the outer surface. Hopefully, this would differentiate between the protein and

DNA portions of the virus. Then, each mixture was spun in a centrifuge to separate the heavy bacteria (with any viral parts that had gone into them) from the liquid solution they were in (including any viral parts that had not entered the bacteria). The centrifuge causes the heavier bacteria to be pulled to the bottom of the tube where they form a pellet, while the light-weight viral "left-overs" stay suspended in the liquid portion called the supernatant.

In the subsequent step, the pellet and supernatant from each tube were separated and tested for the presence of radioactivity. Radioactive sufur was found in the supernatant, indicating that the viral protein did not go into the bacteria. Radioactive phosphorus was found in the bacterial pellet, indicating that viral DNA did go into the bacteria.

Based on these results, Hershey and Chase concluded that DNA must be the genetic code material, not protein as many poeple believed. When their experiment was published and people finally acknowledged that DNA was the genetic material, there was a lot of competition to be the first to discover its chemical structure.

What was known is that DNA contains a nitrogenous base. There are two kinds of these, which include:

Pyrimidine (6-member ring of C & N)	*Purine (that + 5 member ring of C & N)*
Cytosine	Guanine
Thymine in DNAuracil in RNA	Adenine

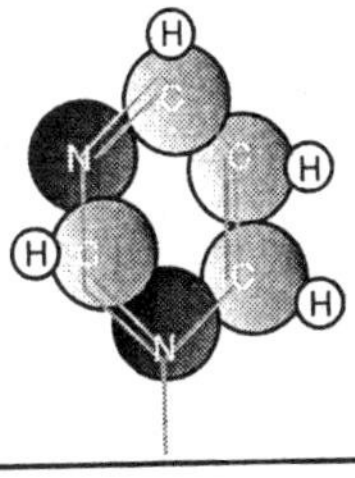

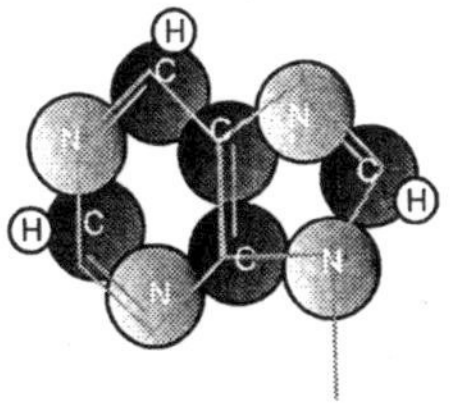

Each nitrogenous base is connected to a molecule of ribose

sugar (–1 oxygen in DNA) to form a nucleoside like the *adenosine* in ATP. Each nucleoside is joined to a PO_4 (phosphate group) to form a nucleotide like adenosine monophosphate (which can be turned into ATP by adding phosphate groups). People also knew that nucleotides were somehow linked by dehydration synthesis to form DNA, but the exact structure/arrangement was unknown.

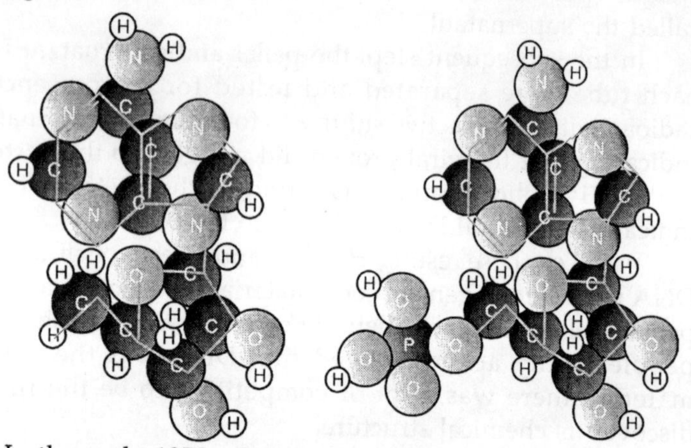

In the early 1950s, Rosalind Franklin, an Englishwoman, was doing research which involved bouncing x-rays off crystals of various substances (a process which is called x-ray crystallography), including DNA, then exposing photographic film to the x-rays. She was studying the scatter patterns made by the x-rays bouncing off the crystals of various substances (Unfortunately, she died of cancer soon afterwards, or she might have been more famous). Other people like Linus Pauling were also attempting to figure out the structure of DNA.

James Watson, a young American scientist was in England working with Francis Crick, another young researcher. Someone else showed them Franklin's photographs of DNA x-ray crystallography, and from her pictures, they were able to determine that the structure of DNA was organized into a double spiral or double helix. Based on Franklin's data, in 1953, Watson and Crick published a paper in which they proposed and described an hypothetical structure for DNA. Subsequent research by many other people has since upheld their hypothesis, and based on subsequent examination of Franklin's lab notes and calculations, she was probably within a couple days of coming to the same conclusion when their paper was published. For their discovery, Watson and Crick received the Nobel prize in 1962. In the intervening time, Rosalind Franklin had died in 1958 of ovarian cancer, probably due in large part to her work with x-rays. Since the Nobel prize is not awarded posthumously, people have often wondered if the Nobel committee would have included Franklin if she had still been alive.

DNA is a double helix. The outer edges are formed of alternating ribose sugar molecules and phosphate groups. The two strands go in opposite directions (1 "up" and 1 "down"). The nitrogenous bases are "inside" like rungs on a ladder. Adenine on one side pairs with thymine (uracil in RNA) on the other by hydrogen bonding, and cytosine pairs with guanine. Note that the C-G pair has three hydrogen bonds while the A-T pair has only two, which keeps them from pairing wrong.

This dictates side-to-side pairing, but says nothing about the order *along* the molecule. Watson and Crick said this variability along the molecule can account for the variety in the genetic code. Their model also accounts for how DNA can replicate itself. They said the molecule "unzips" and new matching bases are added in to create two new molecules. They called this semiconservative replication because each new molecule has one "old" and one "new" strand of DNA.

DNA codes for protein synthesis by first coding for RNA. First, the DNA code is transcribed to RNA code, which is still in the "language" of nitrogenous bases, except that adenine on the DNA pairs with uracil (in place of thymine) on the RNA. The RNA code is then translated to protein code, which is a different "language."

This process involves ribosomes and two kinds of RNA: mRNA and tRNA. The mRNA codes for the gene in question and is copied off the DNA, while tRNA matches a specific group of nucleotides with a specific amino acid. A "unit" of three nucleotides on the tRNA codes for one amino acid. Each of these "units" is called an anticodon. These match up with corresponding three-nucleotide sequences on the mRNA called codons, and in this manner the amino acids are organized into the correct sequence to build a protein. The ribosome works with the mRNA and tRNA to hook the amino acids together to form a protein.

Here is a list of the mRNA codons and the corresponding amino acids for which they code.

		Second Base					
		U	C	A	G		
First Base	U	UUU Phe	UCU Ser	UAU Tyr	UGU Cys	U	**Third Base**
		UUC Phe	UCC Ser	UAC Try	UGC Cys	C	
		UUA Leu	UCA Ser	UAA Stop	UGA Stop	A	
		UUG Leu	UCG Ser	UAG Stop	UGG Trp	G	
	C	CUU Leu	CCU Pro	CAU His	CGU Arg	U	
		CUC Leu	CCC Pro	CAC His	CGC Arg	C	
		CUA Leu	CCA Pro	CAA Gln	CGA Arg	A	
		CUG Leu	CCG Pro	CAG Gln	CGG Arg	G	
	A	AUU Ile	ACU Thr	AAU Asn	AGU Ser	U	
		AUC Ile	ACC Thr	AAC Asn	AGC Ser	C	
		AUA Ile	ACA Thr	AAA Lys	AGA Arg	A	
		AUG Met or Start	ACG Thr	AAG Lys	AGG Arg	G	
	G	GUU Val	GCU Ala	GAU Asp	GGU Gly	U	
		GUC Val	GCC Ala	GAC Asp	GGC Gly	C	
		GUA Val	GCA Ala	GAA Glu	GGA Gly	A	
		GUG Val	GCG Ala	GAG Glu	GGG Gly	G	

Mutations can be caused by a change in the sequence of the nucleotides. Some mutations have more effect than others, depending on where in the code they are and how important that area is to the code. While mutations in some areas of some genes have little effect, sickle cell anemia is caused by a mutation in only one nucleotide. This changes the codon at that location to code for a different amino acid, and that, in turn, significantly changes the shape of the hemoglobin molecules in that person's blood.

When some viruses (especially *Herpes* viruses, including Chicken Pox and Cold Sores) infect us, they insert their DNA into our cells' DNA, and stay resident in our cells for the rest of our lives. These can potentially become active again either making a person sick again (like Shingles in a person who has had Chicken Pox) or just being shed from a person's body (to infect others) without obvious symptoms of illness (like Mononucleosis). Some kinds of cancer may be caused this way. For example, there is some pretty strong evidence linking genital warts (human papillomavirus, HPV) and cervical cancer.

The AIDS virus does things "backwards." This virus contains RNA rather than DNA, yet when it gets into someone's cells, it can do reverse transcription and code from its RNA to make DNA which, then, can code to make more virus.

GENETIC ENGINEERING

We now have the knowledge and ability to transfer genes from one organism to another, which seems to have some benefits associated with it, but may also have many yet-to-be-discovered problems associated with it. Because this is all so new, not enough time has elapsed to allow scientists to study/look for any possible long-term effects of genetically-modified organisms (GMOs).

Many medicines are now made by GMOs. For example, insulin was formerly extracted from the pancreas of animals after they were slaughtered for meat. However, now most insulin is produced by bacteria with the insulin gene spliced onto their chromosome. Using this method of production, drug companies can make more insulin, faster.

In theory, insulin produced in this way should be more "pure" — someone who could not use pig insulin due to a pork allergy may be able to tolerate insulin made in this way (but someone could, potentially, be allergic to some component of the bacteria present in the refined insulin). In the relatively short time that this form of insulin has been available, there is no doubt that it has saved many people's lives — let's hope that some unforseen, long-term, deleterious effects are not discovered later on.

Experimentation is being done to investigate the possible use of genetically-engineered viruses to treat genetic diseases such as cystic fibrosis. In this "treatment," a kind of virus that infects our lungs is used. The genes that enable it to infect our cells are kept while the genes which make us sick are (hopefully) all removed, and the missing human (normal) gene that relieves cystic fibrosis is then inserted into the virus' genome. These viruses are then sprayed into the lungs of a person with CF and allowed to "infect" the cells in that person's lungs.

When the normal gene is inserted into the genetic make-up of the cells lining that person's lungs, those cells function normally and the CF symptoms are alleviated. However, this treatment doesn't last because only that layer of cells is "infected," and when those cells die and are replaced by new cells, the new cells do not contain the genetic code to overcome the CF gene, and the person must inhale more genetically-engineered virus. While this seems to be a promising, life-giving, technique, no data are yet available on long-term effects and safety.

Bacillus thuringiensis (Bt) is a species of bacterium that infects and kills a number of species of caterpillars (Remember that caterpillars turn into butterflies or moths when they grow up.). There is a species of moth whose caterpillar is a "pest" in corn plants, and insects of any kind are more successful when we humans plant huge monocultures of their favourite foods. For a number of years, now, people have realized that in certain situations, by judiciously applying BT, "pest" species of caterpillars can be infected and killed without the use of man-made, chemical insecticides.

More recently, a major US chemical company came up with the idea of creating genetically-engineered corn containing BT genes, which was good news to agri-business firms who plant huge areas of land with monocultures of corn. Because corn is wind-pollinated and therefore makes lots of pollen which, in this case, contains BT genes, many scientists are concerned about the effects of this corn on local butterfly populations.

Some research has indicated that when this pollen settles on nearby caterpillar host plants and is, therefore, consumed by caterpillars, this might cause an increase in mortality (therefore fewer "good" butterflies such as Monarchs). It is assumed that these BT genes in corn should, they think, have no adverse effects on people or cattle who consume this corn, but no long-term testing has been done. Also, this company has convinced the government that this corn should be marketed without any labeling indicating that it is genetically-engineered — they're scared that if you are given a choice, you won't buy/eat their corn if you know it is genetically engineered.

Additionally, to increase their profits, this company has also put "suicide" genes into this corn so that farmers cannot save seed from one year to plant the next year, and have to buy more seed from them, instead. For big agri-businesses, this is of small consequence, but for small, family farmers this is total disaster. To save money, the latter often save seed from one year to plant the next, and even if they don't plant this genetically-engineered corn, if their corn is pollinated by genetically-engineered corn from a neighboring field, it will not produce viable seed.

This same chemical company came up with the idea of "transplanting" what they think is the cold-tolerance gene from a species of cold-water-inhabiting fish into tomatoes, thereby hoping to "invent" cold-tolerant tomatoes, and again, has convinced the government that these tomatoes should not be labeled in any way to indicate that they have fish genes in them, and that you should not have a choice about what you eat.

At the very least, this would be a problem for someone who is a vegetarian and chooses to not eat fish. A more serious consequence, however, would be that a person who is severely allergic to fish could also have an allergic reaction to these tomatoes and end up in the hospital. Again, no tests were done before the government approved these tomatoes and no data are available on the long-term effects on humans (or anything else).

This same chemical company manufactures a widely-used herbicide, and came up with the idea to genetically engineer cotton, soybeans, and other crop plants so they would be immune to the effects of that herbicide. That way, farmers can (have to?) buy their seed from that company, then spray their fields with herbicide also purchased from that company to (hopefully) kill all local plants except the immune crop, simultaneously putting all their money in the corporation's pockets.

Preliminary research has shown that these resistant genes have already begun to "jump" into local weeds, thereby making them resistant to that herbicide, and again, no data are available on long-term effects on humans, other animals, or the environment in general.

DISCOVERY OF THE STRUCTURE OF DNA

Most biological experiments are done with samples of living matter- cells, tissues, whole organisms or extracts of these materials. Only a few experiments are done with mathematical or physical models of biological components. Ocassionally, however, a model experiment gives an insight that would be difficult to obtain in any other way. This was true of the discovery of the structure of DNA. A crucial experiment was done by James Watson in 1952 using nothing more complicated than chemical models cut out of cardboard. The experiment is so simple that you can do it yourself and can even show it to your children and friends.

PREVIOUS INFORMATION ABOUT THE STRUCTURE OF DNA

By 1952 conventional laboratory experiments had shown the following:

- DNA was the molecule of heredity: it had been shown that transferring DNA into bacteria could change them genetically. This made the solving of the DNA structure one of the top priorities in biology
- DNA was known to be composed of phosphate, the 5 carbon sugar deoxyribose and 4 nucleotide bases: adenine, cytosine, guanine and thymine (abbreviated A, C, G & T)

- Chemical structures of all of the components (phosphate, sugar & bases) were known
- The sugar (S) and phosphate (P) were known to be connected together to form a backbone. The bases (B) were known to be stuck out to the side, attached to the sugar of the backbone by one of their N atoms:

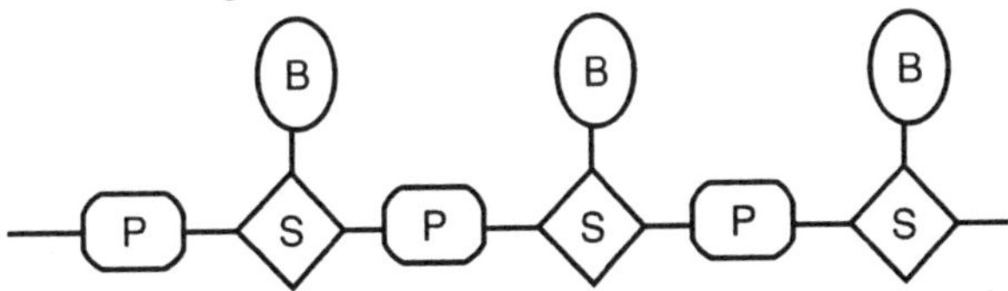

- Rosalind Franklin had pointed out that phosphate has negative charges at cellular pHs. Since these charges would repel each other she insisted that the sugar-phosphate backbone was on the outside of the molecule
- Hydrogen bonds had been shown to be very important in determining the shapes of proteins and it was expected that they would be important in DNA as well
- Rosalind Franklin had taken excellent x-ray diffraction pictures of DNA. The pictures had a distinctive pattern that was known from theory (developed by Francis Crick) to be due to a helical structure (2 or more strands spiraling around each other)
- Erwin Chargaff had been making accurate measurements of the base composition of DNA from many different species. He had observed a very strange relationship: in all cases A = T and C = G

Almost everything needed to solve the structure of DNA is in this list, but although a number of people had this data no one had come up with a reasonable solution (several "unreasonable" solutions had been suggested and even published).

AN EXPERIMENT WITH CARDBOARD MODELS

In late 1952 several laboratories were actively working on the structure of DNA. Linus Pauling at Cal Tech had published an incorrect structure with 3 DNA chains wound in a helix. Rosalind Franklin and Maurice Wilkins were doing x-ray diffraction experiments at King's College in London. Erwin Chargaff thought that the secret to the structure was in the base composition of DNA and he was doing painstaking chemical analyses at Columbia University.

Finally, at Cambridge University Francis Crick and a young American postdoctoral student, James Watson, were very interested in the DNA structure. They were an unlikely pair because they were doing no laboratory experiments of their own. Instead they were talking to all of the other participants and building chemical models. After a number of mistakes they ordered a set a metal models of the 4 bases to be made by the university shop.

The models were slow in coming, so one afternoon Watson became impatient and decided to draw the base structures on cardboard and cut them out. He did this and then went out to dinner and the theatre.

The next morning he returned to his models and moved them around in pairs to see how they might fit together. He was starting to think that the DNA had 2 chains and he wanted to see if interactions between the bases might hold them together.

It was immediately apparent that interactions between the bases would involve hydrogen bonds. Within a few minutes the structure of DNA became apparent and Watson had an explanation for Chargaff's results (A = T and C = G) and even a good idea of how DNA replicated itself.

The objective is to move the models around to see how they might fit together. You must pay special attention to possible hydrogen bonds. Hydrogen bonds occur between 2 groups of atoms:

- Hydrogen atoms attached to oxygen and nitrogen
- Hydrogen acceptors: oxygens and nitrogens without attached hydrogens

In the drawings of the A, C, G and T bases below the hydrogens available for H bonds are colored red and the hydrogen acceptors are colored green, as shown in this table:

C	Carbon atoms (don't make good H bonds)
N	Nitrogen atoms attached to deoxyribose sugar (not available for H bonding).
N	Nitrogen atoms already bonded to H (not available for H bonding).
N O	Nitrogen & Oxygen atoms available for H bonding
	H atoms attached to Carbon (don't make good H bonds)
	H atoms available for H bonding

Here are the 4 DNA bases with the correct colour-coding:

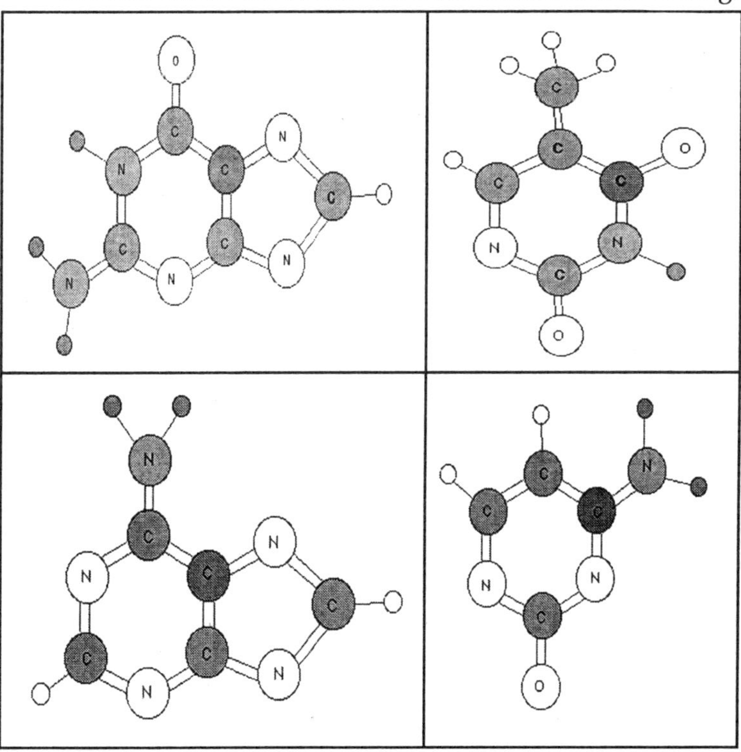

DNA AND MOLECULAR GENETICS

THE PHYSICAL CARRIER OF INHERITANCE

While the period from the early 1900s to World War II has been considered the "golden age" of genetics, scientists still had not determined that DNA, and not protein, was the hereditary material. However, during this time a great many genetic discoveries were made and the link between genetics and evolution was made. Friedrich Meischer in 1869 isolated DNA from fish sperm and the pus of open wounds. Since it came from nuclei, Meischer named this new chemical, nuclein. Subsequently the name was changed to nucleic acid and lastly to deoxyribonucleic acid (DNA).

Robert Feulgen, in 1914, discovered that fuchsin dye stained DNA. DNA was then found in the nucleus of all eukaryotic cells. During the 1920s, biochemist P.A. Levene analyzed the components of the DNA molecule. He found it contained four nitrogenous bases: cytosine, thymine, adenine, and guanine; deoxyribose sugar; and a phosphate group. He concluded that the basic unit (nucleotide) was composed of a base attached to a sugar and that the phosphate also attached to the sugar. He (unfortunately) also erroneously concluded that the proportions of bases were equal and that there was a tetranucleotide that was the repeating structure of the molecule.

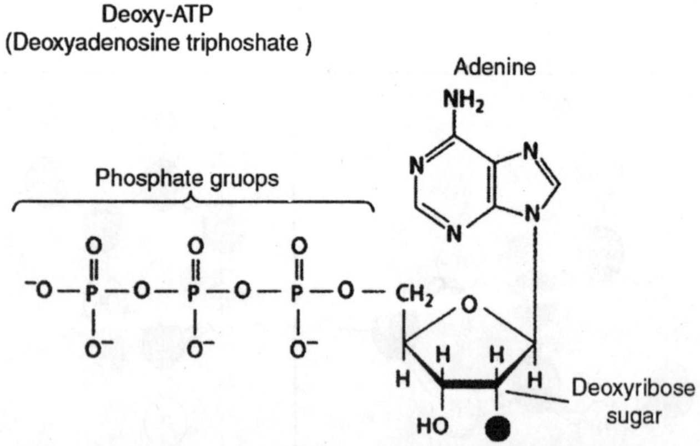

Deoxy-ATP
(Deoxyadenosine triphoshate)

Adenine

Phosphate gruops

Deoxyribose sugar

Fig. Molecular Structure of Three Nirogenous Bases

During the early 1900s, the study of genetics began in earnest: the link between Mendel's work and that of cell biologists resulted in the chromosomal theory of inheritance; Garrod proposed the link between genes and "inborn errors of metabolism"; and the question was formed: what is a gene?

The answer came from the study of a deadly infectious disease: *pneumonia*. During the 1920s Frederick Griffith studied the difference between a disease-causing strain of the pneumonia causing bacteria (*Streptococcus peumoniae*) and a strain that did not cause pneumonia.

The pneumonia-causing strain (the S strain) was surrounded by a capsule. The other strain (the R strain) did not have a capsule and also did not cause pneumonia. Frederick Griffith was able to induce a nonpathogenic strain of the bacterium *Streptococcus pneumoniae* to become pathogenic. Griffith referred to a transforming factor that caused the non-pathogenic bacteria to become pathogenic.

Griffith injected the different strains of bacteria into mice. The S strain killed the mice; the R strain did not. He further noted that if heat killed S strain was injected into a mouse, it did not cause pneumonia.

When he combined heat-killed S with Live R and injected the mixture into a mouse (remember neither alone will kill the mouse) that the mouse developed pneumonia and died. Bacteria recovered from the mouse had a capsule and killed other mice when injected into them!

Hypotheses:

- The dead S strain had been reanimated/resurrected.
- The Live R had been transformed into Live S by some "transforming factor".

Further experiments led Griffith to conclude that number 2 was correct.

In 1944, Oswald Avery, Colin MacLeod, and Maclyn McCarty revisited Griffith's experiment and concluded the transforming factor was DNA. Their evidence was strong but not totally conclusive. The then-current favourite for the hereditary material was protein; DNA was not considered by many scientists to be a strong candidate.

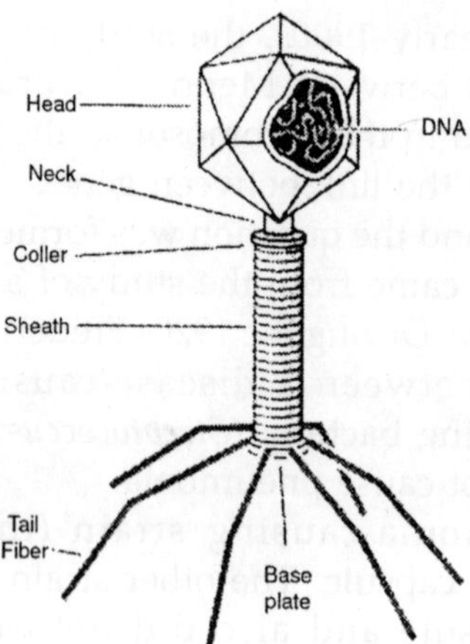

Fig. Structure of a Bacteriophage Virus

The breakthrough in the quest to determine the hereditary material came from the work of Max Delbruck and Salvador Luria in the 1940s. Bacteriophage are a type of virus that attacks bacteria, the viruses that Delbruck and Luria worked with were those attacking *Escherichia coli*, a bacterium found in human intestines. Bacteriophages consist of protein coats covering DNA. Bacteriophages infect a cell by injecting DNA into the host cell.

This viral DNA then "disappears" while taking over the bacterial machinery and beginning to make new virus instead of new bacteria. After 25 minutes the host cell bursts, releasing hundreds of new bacteriophage.

Phages have DNA and protein, making them ideal to resolve the nature of the hereditary material. In 1952, Alfred D. Hershey and Martha Chase conducted a series of experiments to determine whether protein or DNA was the hereditary material. By labeling the DNA and protein with different (and mutually exclusive) radioisotopes, they would be able to determine which chemical (DNA or protein) was getting into the bacteria.

Such material must be the hereditary material (Griffith's transforming agent). Since DNA contains Phosphorous (P) but no Sulfur (S), they tagged the DNA with radioactive Phosphorous-32. Conversely, protein lacks P but does have S, thus it could be tagged with radioactive Sulfur-35. Hershey and Chase found that the radioactive S remained outside the cell while the radioactive P was found inside the cell, indicating that DNA was the physical carrier of heredity.

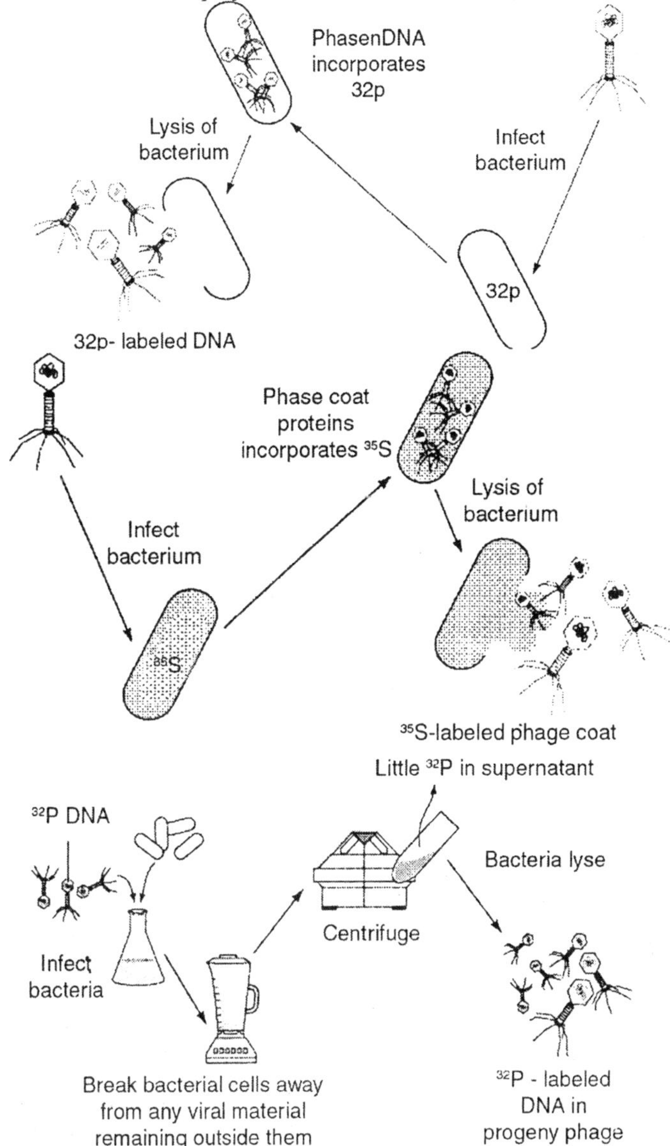

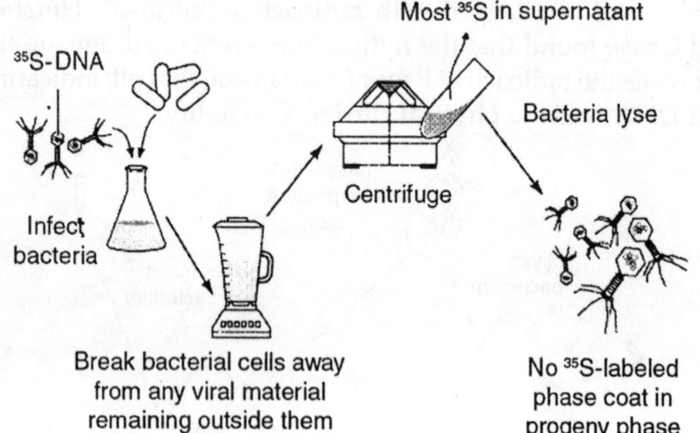

Fig. Diagrams Illlustrating the Hershey and Chase Experiment that Supported DNA as the Hereditary Material while it also showed Protein was NOT the Hereditary Material

THE STRUCTURE OF DNA

Erwin Chargaff analyzed the nitrogenous bases in many different forms of life, concluding that the amount of purines does not always equal the amount of pyrimidines (as proposed by Levene). DNA had been proven as the genetic material by the Hershey-Chase experiments, but how DNA served as genes was not yet certain.

DNA must carry information from parent cell to daughter cell. It must contain information for replicating itself. It must be chemically stable, relatively unchanging. However, it must be capable of mutational change. Without mutations there would be no process of evolution.

Many scientists were interested in deciphering the structure of DNA, among them were Francis Crick, James Watson, Rosalind Franklin, and Maurice Wilkens. Watson and Crick gathered all available data in an attempt to develop a model of DNA structure.

Franklin took X-ray diffraction photomicrographs of crystalline DNA extract, the key to the puzzle. The data known at the time was that DNA was a long molecule, proteins were helically coiled (as determined by the work of Linus Pauling),

Chargaff's base data, and the x-ray diffraction data of Franklin and Wilkens.

DNA is a double helix, with bases to the centre (like rungs on a ladder) and sugar-phosphate units along the sides of the helix (like the sides of a twisted ladder). The strands are complementary (deduced by Watson and Crick from Chargaff's data, A pairs with T and C pairs with G, the pairs held together by hydrogen bonds).

Notice that a double-ringed purine is always bonded to a single ring pyrimidine. Purines are Adenine (A) and Guanine (G). We have encountered Adenosine triphosphate (ATP) before, although in that case the sugar was ribose, whereas in DNA it is deoxyribose.

Pyrimidines are Cytosine (C) and Thymine (T). The bases are complementary, with A on one side of the molecule you only get T on the other side, similarly with G and C. If we know the base sequence of one strand we know its complement.

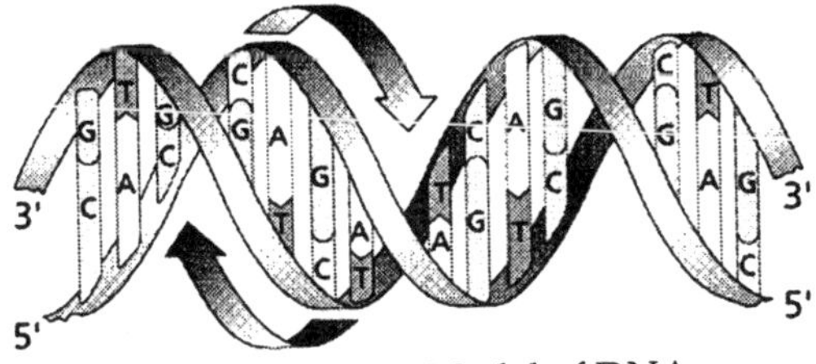

Fig. The Ribbon Model of DNA

DNA REPLICATION

DNA was proven as the hereditary material and Watson et al. had deciphered its structure. What remained was to determine how DNA copied its information and how that was expressed in the phenotype. Matthew Meselson and Franklin W. Stahl designed an experiment to determine the method of DNA replication. Three models of replication were considered likely.

Conservative replication

Would somehow produce an entirely new DNA strand during replication.

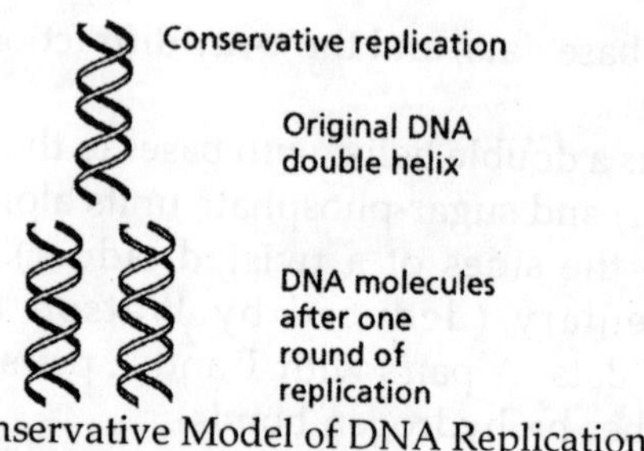

Fig. Conservative Model of DNA Replication

Semiconservative Replication

Would produce two DNA molecules, each of which was composed of one-half of the parental DNA along with an entirely new complementary strand. In other words the new DNA would consist of one new and one old strand of DNA. The existing strands would serve as complementary templates for the new strand.

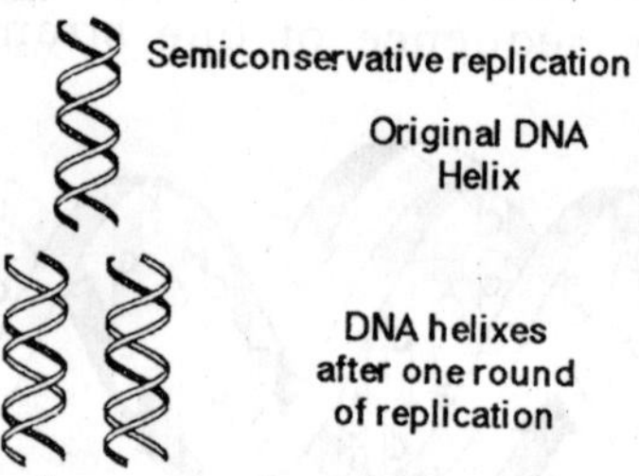

Fig. The Semiconservative Model of DNA Structure

Dispersive replication

Involved the breaking of the parental strands during replication, and somehow, a reassembly of molecules that were a mix of old and new fragments on each strand of DNA.

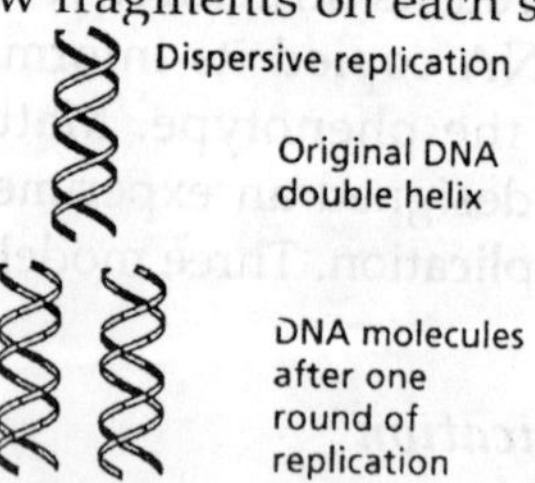

Fig. The Dispersive Replication Model of DNA Replication

The Meselson-Stahl experiment involved the growth of

E. coli bacteria on a growth medium containing heavy nitrogen (Nitrogen-15 as opposed to the more common, but lighter molecular weight isotope, Nitrogen-14). The first generation of bacteria was grown on a medium where the sole source of N was Nitrogen-15. The bacteria were then transferred to a medium with light (Nitrogen-14) medium. Watson and Crick had predicted that DNA replication was semi-conservative. If it was, then the DNA produced by bacteria grown on light medium would be intermediate between heavy and light.

It was. DNA replication involves a great many building blocks, enzymes and a great deal of ATP energy (remember that after the S phase of the cell cycle cells have a G phase to regenerate energy for cell division). Only occurring in a cell once per (cell) generation, DNA replication in humans occurs at a rate of 50 nucleotides per second, 500/second in prokaryotes. Nucleotides have to be assembled and available in the nucleus, along with energy to make bonds between nucleotides.

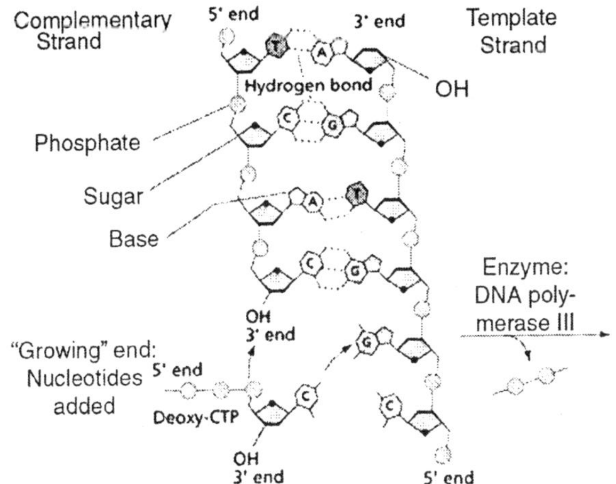

DNA polymerases unzip the helix by breaking the H-bonds between bases. Once the polymerases have opened the molecule, an area known as the replication bubble forms (always initiated at a certain set of nucleotides, the origin of replication). New nucleotides are placed in the fork and link to the corresponding parental nucleotide already there (A with

T, C with G). Prokaryotes open a single replication bubble, while eukaryotes have multiple bubbles. The entire length of the DNA molecule is replicated as the bubbles meet.

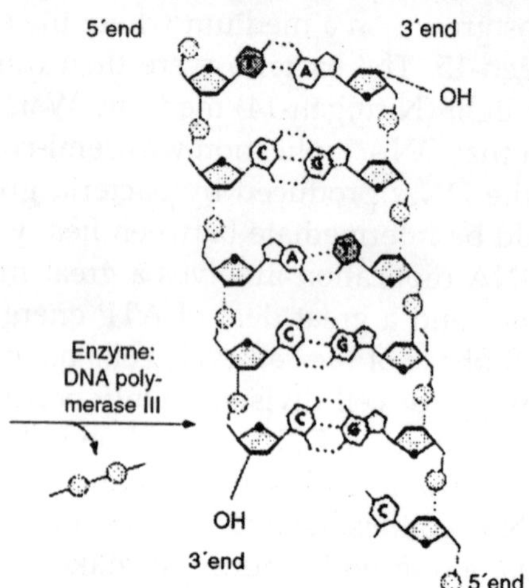

Fig. The Roles of DNA Polymerases in

Since the DNA strands are antiparallel, and replication proceeds in thje 5' to 3' direction on EACH strand, one strand will form a continuous copy, while the other will form a series of short Okazaki fragments.

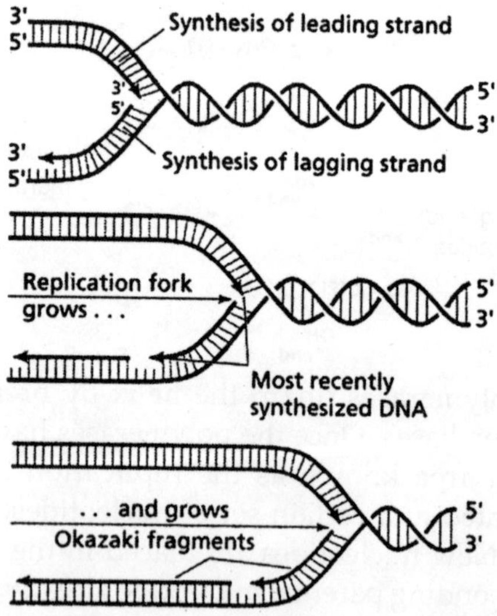

ENZYME STRUCTURE PROVIDES CLUES TO DNA

Before a cell can begin to divide or differentiate, the genetic information within the cell's DNA must be copied, or "transcribed," onto complementary strands of RNA. RNA polymerase II (pol II) is an enzyme that, by itself, can unwind the DNA double helix, synthesize RNA, and proofread the result. When combined with other molecules that regulate and control the transcription process, pol II is the key to successful interpretion of an organism's genetic code.

However, the size, complexity, scarcity, and fragility of pol II complexes have made analysis of these macromolecules by x-ray crystallography a formidable challenge. A team of structural biologists has met this challenge using data obtained from both the Stanford Synchrotron Radiation Laboratory and the Macromolecular Crystallography Facility at the ALS. The resultant high-resolution model of a 10-subunit pol II complex suggests roles for each of the subunits and will allow researchers to begin unraveling the intricacies of DNA transcription and its role in gene expression. In this work, the researchers studied the pol II enzyme from the yeast *Saccharomyces cerevisiae*, which is likely to be an excellent model for the human enzyme in light of its highly similar gene sequences.

It is also the best-characterized form of the pol II enzyme, having been the subject of many biochemical and low-resolution structural studies in the past. To obtain a high-resolution structure, the research team drew on its considerable expertise in the preparation of protein crystals: two-dimensional crystals of pol II (minus two small subunits found to impede crystal growth) were used as seeds for growing three-dimensional crystals. These crystals, when produced in an inert atmosphere to prevent oxidation, enabled the collection of data to 3.5-angstrom resolution. The addition of a final soaking procedure to produce uniform crystals, combined with high-brightness x-ray sources, resulted in a resolution of 3.0 angstroms.

The current results bring into focus the somewhat fuzzy features previously observed in or inferred from earlier experiments. More importantly, the structural details suggest possible explanations for some of the unusual characteristics of this enzyme, which include a high processivity (the ability to synthesize very long strands of RNA) and the tendency to work in periodic spurts separated by pauses.

While it is known that additional proteins (transcription factors) play a role in controlling the activity of pol II, scientists have yet to understand how such proteins interact with pol II binding sites to perform their various functions. The pol II model reported here establishes the positions of the various subunits and provides detailed information about the DNA/RNA binding domains.

The data reveal two main subunits (Rpb1 and Rpb2) separated by a deep cleft where DNA can enter the complex. At the end of the cleft is the active site, where the DNA can be unwound for a short distance (the "transcription bubble") and a DNA/RNA hybrid can be produced. Two prominent grooves lead away from the active site, either of which could accommodate the exiting RNA transcript. An opening below the active site may allow the entry of nucleotides (for manufacturing RNA) and transcription factors (for regulating the process).

The same opening may provide room for the leading end of the RNA strand during "backtracking" maneuvers, which are important for proofreading and for traversing obstacles such as DNA damage.

Other notable features that might help account for the great stability of this transcribing complex include a pair of "jaws" that appear to grip the DNA strands as they enter the complex and, closer to the active site, a clamp on the DNA that could possibly be locked in the closed position by the presence of RNA.

The high-resolution pol II structure reported here is a landmark achievement, pulling together threads from numerous diverse research efforts into a cohesive whole. Further study should yield many new insights into the detailed mechanisms of pol II and its transcription factors. Construction of an atomic model is already well underway.

UNRAVELING DNA

Encoded into the double-helical strands of DNA, the human genome is the complete set of instructions required to make a human being. While the mapping of the human genome (roughly three billion components) is certainly a Herculean accomplishment, it is only the first step toward realizing the full potential of genomic medicine in the diagnosis, monitoring, and treatment of disease. Beyond knowing what the genetic blueprint says, scientists must understand how that blueprint gets interpreted, or "expressed" as an individual with unique traits. The pol II enzyme is the catalyst for a major step in this process.

As a pol II molecule slides along a DNA molecule, it "unzips" the strands of the DNA double helix, synthesizes a complementary strand of RNA (which will carry the genetic information to where it is needed), and verifies that no mistakes have occurred.

This process is regulated by transcription factors—separate molecules that bind to pol II and determine which genes are expressed, at what stage of development, and in which tissue.

Done correctly, this process results in healthy cell growth and differentiation; otherwise, aberrations such as cancer can be the result. Thus, details of the structure of pol II, including information about its binding sites and how they interact with transcription factors, will provide valuable insight into the detailed mechanisms underlying the flow of genetic information from DNA to RNA to protein, which is necessary for life and health.

HYDROGEN BONDING IN DNA

Without hydrogen bonds there could be no life because they hold the double helix of DNA together, and this they do by charge attractions. A hydrogen bonded to oxygen, or nitrogen, becomes slightly positively charged which enables it to attract a centre of negative charge on another molecule, such as another oxygen or nitrogen atom. The hydrogen bond is then written, e.g., O-H...N, with the dotted line signifying the hydrogen bond. There are also O-H...O, N-H...O and N-H...N bonds, the last being among the weakest.

The secondary effects they have on structures, molecular vibrations, etc., can be used to infer hydrogen bonding, but there is no primary way of observing them because of their inherent weakness. NMR appears to be the least useful technique because neither of the common isotopes, oxygen-16 or nitrogen-14, has a magnetic nucleus. However, nitrogen-15 has a magnetic moment, and by replacing ^{14}N by N, Grzesiek has opened up a new area of investigating these enigmatic bonds.

Paper reports for the first time the direct observation by NMR of an N-H...N hydrogen bond between nucleic acids enriched with ^{15}N, by measuring the coupling of the nitrogen atoms. Grzesiek, working with Andrew Dingley of the Heinrich-Heine University in Düsseldorf, has been able to do this and show that the coupling is surprisingly large. Normally atomic nuclei only couple with each other if they are linked by normal chemical bonds, and in theory hydrogen bonds have neither the strength nor stability for this to occur.

The German researchers studied an ^{15}N enriched sample of the T1 domain of the potato spindle tuber viroid and were able to prove that N-H...N hydrogen bonding was present between the base pairs, uridine...adenosine and guanosine...cytidine, with couplings of approximately 7 Hz. How could they be certain that the signals they were observing are due to N-H...N hydrogen bonds? The answer was to use triple resonance techniques to examine base pairs that hydrogen bond only via O-H...N hydrogen bonds, and show that the signal they had previously observed was absent.

Grzesiek's second paper on hydrogen bonds, coauthored by Florence Cordier, Heinrich-Heine University, extends the work in an even more remarkable way by measuring the NMR coupling between nitrogens and carbons in the backbone hydrogen bonds of the human protein ubiquitin. The carbons are part of a carbonyl (C=O) group, so are one removed from the hydrogen bond, i.e., N-H...O=C.

This time they used material enriched with ^{15}N and ^{13}C

(normal ^{12}C has no nuclear magnet), and there too was the evidence for these hydrogen bonds, albeit with an interaction an order of magnitude weaker (at -0.25 to -0.9 Hz) than the N-H...N coupling. Nevertheless, the couplings correlate with the strength of the hydrogen bond, being stronger in the stronger bonds. These findings were confirmed by paper #9, which is from the researchers of Ad Bax's group based at NIH, Bethesda, Maryland.

Grzesiek's third paper, was done in conjunction with Dingley and researchers at UCLA. Together they studied not only the hydrogen bonding of Watson-Crick base pairs but also of Hoogsten base pairs within a DNA triplex consisting of one purine and two pyrimidine strands. Four different base pairs were identified, their various couplings distinguished–including those of the weaker interactions at the "frayed ends" of the DNA chains–and relationships with other hydrogen bonding parameters, such as bond length, were established. In addition they were able to show that density functional computer simulations by computer could reproduce these findings exactly.

Chapter 3

RNA Structure

We now turn our attention to RNA, which differs from DNA in three respects. First, the backbone of RNA contains ribose rather than 2′-deoxyribose. That is, ribose has a hydroxyl group at the 2′ position. Second, RNA contains uracil in place of thymine. Uracil has the same single-ringed structure as thymine, except that it lacks the 5′ methyl group. Thymine is in effect 5′methyl-uracil. Third, RNA is usually found as a single polynucleotide chain.

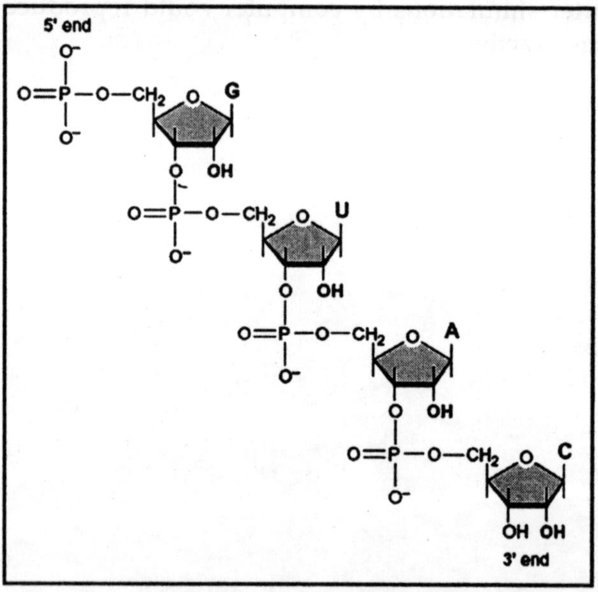

Fig. Structural Features of RNA

Except for the case of certain viruses, RNA is not the

genetic material and does not need to be capable of serving as a template for its own replication. Rather, RNA functions as the intermediate, the mRNA, between the gene and the protein-synthesizing machinery. Another function of RNA is as an adaptor, the tRNA, between the codons in the mRNA and amino acids. RNA can also play a structural role as in the case of the RNA components of the ribosome.

Yet another role for RNA is as a regulatory molecule, which through sequence complementarity binds to, and interferes with the translation of, certain mRNAs. Finally, some RNAs (including one of the structural RNAs of the ribosome) are enzymes that catalyze essential reactions in the cell. In all of these cases, the RNA is copied as a single strand off only one of the two strands of the DNA template, and its complementary strand does not exist. RNA is capable of forming long double helices, but these are unusual in nature.

RNA CHAINS FOLD BACK ON THEMSELVES TO FORM LOCAL REGIONS

Despite being single-stranded, RNA molecules often exhibit a great deal of double-helical character. This is because RNA chains frequently fold back on themselves to form base-paired segments between short stretches of complementary sequences. If the two stretches of complementary sequence are near each other, the RNA may adopt one of various stem-loop structures in which the intervening RNA is looped out from the end of the double-helical segment as in a hairpin, a bulge, or a simple loop. The stability of such stem-loop structures is in some instances enhanced by the special properties of the loop.

For example, a stem-loop with the "tetraloop" sequence UUCG is unexpectedly stable due to special base-stacking interactions in the loop. Base pairing can also take place between sequences that are not contiguous to form complex structures aptly named pseudoknots. The regions of base pairing in RNA can be a regular double helix or they can contain discontinuities, such as noncomplementary nucleotides that bulge out from the helix.

A feature of RNA that adds to its propensity to form double-helical structures is an additional, non-Watson-Crick base pair. This is the G:U base pair, which has hydrogen bonds between N3 of uracil and the carbonyl on C6 of guanine and between the carbonyl on C2 of uracil and N1 of guanine. Because G:U base pairs can occur as well as the four conventional, Watson-Crick base pairs, RNA chains have an enhanced capacity for self-complementarity. Thus, RNA frequently exhibits local regions of base pairing but not the long-range, regular helicity of DNA. The presence of 2'-hydroxyls in the RNA backbone prevents RNA from adopting a B-form helix. Rather, double-helical RNA resembles the A-form structure of DNA. As such, the minor groove is wide and shallow, and hence accessible, but recall that the minor groove offers little sequence-specific information. Meanwhile, the major groove is so narrow and deep that it is not very accessible to amino acid side chains from interacting proteins. Thus, the RNA double helix is quite distinct from the DNA double helix in its detailed atomic structure and less well suited for sequence-specific interactions with proteins (although some proteins do bind to RNA in a sequence-specific manner).

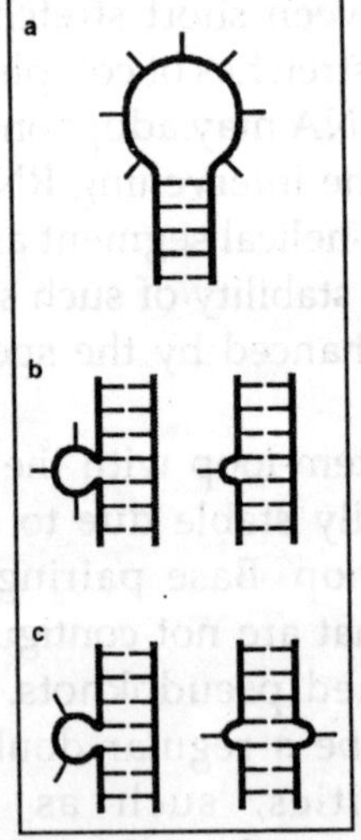

Fig. Double Helical Characteristics of RNA.

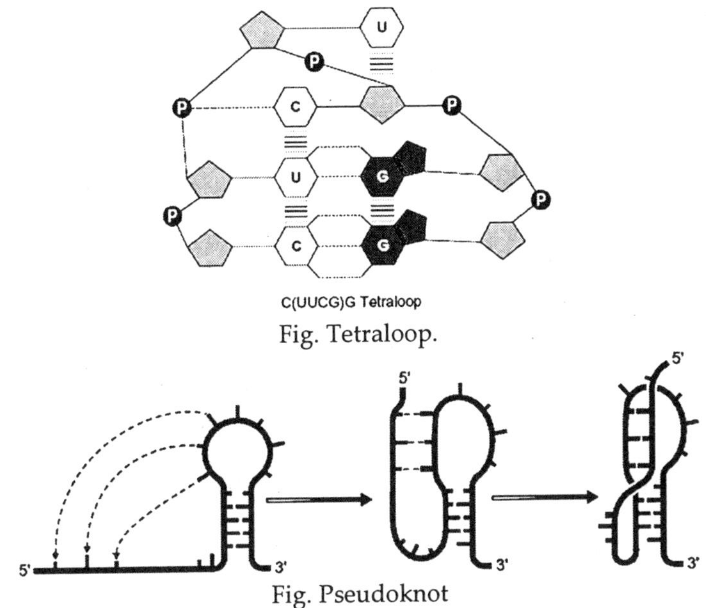

Fig. Tetraloop.

Fig. Pseudoknot

RNA CAN FOLD UP INTO COMPLEX TERTIARY STRUCTURES

Freed of the constraint of forming long-range regular helices, RNA can adopt a wealth of tertiary structures. This is because RNA has enormous rotational freedom in the backbone of its non-base-paired regions. Thus, RNA can fold up into complex tertiary structures frequently involving unconventional base pairing, such as the base triples and base-backbone interactions seen in tRNAs. Proteins can assist the formation of tertiary structures by large RNA molecules, such as those found in the ribosome.

Proteins shield the negative charges of backbone phosphates, whose electrostatic repulsive forces would otherwise destabilize the structure. Researchers have taken advantage of the potential structural complexity of RNA to generate novel RNA species (not found in nature) that have specific desirable properties. By synthesizing RNA molecules with randomized sequences, it is possible to generate mixtures of oligonucleotides representing enormous sequence diversity.

For example, a mixture of oligoribonucleotides of length 20 and having four possible nucleotides at each position would have a potential complexity of 420 sequences or 1012 sequences! From mixtures of diverse oligoribonucleotides, RNA molecules can be selected biochemically that have particular properties, such as an affinity for a specific small molecule.

Some RNAs Are Enzymes

It was widely believed for many years that only proteins could be enzymes. An enzyme must be able to bind a substrate, carry out a chemical reaction, release the product and repeat this sequence of events many times. Proteins are well suited to this task because they are composed of many different kinds of amino acids and they can fold into complex tertiary structures with binding pockets for the substrate and small molecule cofactors and an active site for catalysis.

Now we know that RNAs, which as we have seen can similarly adopt complex tertiary structures, can also be biological catalysts. Such RNA enzymes are known as ribozymes, and they exhibit many of the features of a classical enzyme, such as an active site, a binding site for a substrate and a binding site for a cofactor, such as a metal ion. One of the first ribozymes to be discovered was RNase P, a ribonuclease that is involved in generating tRNA molecules from larger, precursor RNAs. RNase P is composed of both RNA and protein; however, the RNA moiety alone is the catalyst. The protein moiety of RNase P facilitates the reaction by shielding the negative charges on the RNA so that it can bind effectively to its negatively charged substrate. The RNA moiety is able to catalyze cleavage of the tRNA precursor in the absence of the protein if a small, positively charged counter ion, such as the peptide spermidine, is used to shield the repulsive, negative charges. Other ribozymes carry out trans-esterification reactions involved in the removal of intervening sequences known as introns from precursors to certain mRNAs, tRNAs, and ribosomal RNAs in a process known as RNA splicing.

THE HAMMERHEAD RIBOZYME

Before concluding our discussion of RNA, let us look in more detail at the structure and function of one particular ribozyme, the hammerhead. The hammerhead is a sequence-specific ribonuclease that is found in certain infectious RNA agents of plants known as *viroids,* which depend on self-cleavage to propagate. When the viroid replicates, it produces multiple copies of itself in one continuous RNA chain. Single viroids arise by cleavage, and this cleavage reaction is carried out by the RNA sequence around the junction.

One such self-cleaving sequence is called the *hammerhead* because of the shape of its secondary structure, which consists of three base-paired stems surrounding a core of non-complementary nucleotides required for catalysis. The tertiary structure of the ribozyme, however, looks more like a wishbone. To understand how the hammerhead works, let us first look at how RNA undergoes hydrolysis under alkaline conditions.

At high pH, the 2′ hydroxyl of the ribose in the RNA backbone can become deprotonated, and the resulting negatively charged oxygen can attack the scissile phosphate at the 3′ position of the same ribose. This reaction breaks the RNA chain, producing a 2′, 3′ cyclic phosphate and a free 5′ hydroxyl. Each ribose in an RNA chain can undergo this reaction, completely cleaving the parent molecule into nucleotides. Many protein ribonucleases also cleave their RNA substrates via the formation of a 2′, 3′ cyclic phosphate. Working at normal cellular pH, these protein enzymes use a metal ion, bound at their active site, to activate the 2′ hydroxyl of the RNA. The hammerhead is a sequencespecific ribonuclease, but it too cleaves RNA via the formation of a 2′, 3′ cyclic phosphate.

Hammerhead-mediated cleavage involves a ribozyme-bound Mg″ ion that deprotonates the 2′ hydroxyl at neutral pH, resulting in nucleophilic attack on the scissile phosphate. Because the normal reaction of the hammerhead is self-cleavage, it is not really a catalyst; each molecule normally promotes a reaction one time only, thus having a turnover number of one. But the hammerhead can be engineered to function as a true ribozyme by dividing the molecule into two portions—one, the ribozyme, that contains the catalytic core and the other, the substrate, that contains the cleavage site. The substrate binds to the ribozyme at stems I and III. After cleavage, the substrate is released and replaced by a fresh uncut substrate, thereby allowing repeated rounds of cleavage.

LIFE EVOLVE FROM AN RNA

The discovery of ribozymes has profoundly altered our view of how life might have evolved. We can now imagine that there was a primitive form of life based entirely on RNA. In this world, RNA would have functioned as the genetic material and as the enzymatic machines. This RNA world would have preceded life as we know it today, in which information transfer is based on DNA, RNA, and protein.

A hint that the protein world might have arisen from an RNA world is the discovery that the component in the ribosome that is responsible for the formation of the peptide bond, the peptidyl transferase, is an RNA molecule. Unlike RNase P, the hammerhead, and other previously known ribozymes which act on phosphorous centers, the peptidyl transferase acts on a carbon centre to create the peptide bond.

It thus links RNA chemistry to the most fundamental reaction in the protein world, peptide bond formation. Perhaps then the ribosome ribozyme is a relic of an earlier form of life in which all enzymes were RNAs. DNA is usually in the form of a right-handed double helix. The helix consists of two polydeoxynucleotide chains. Each chain is an alternating polymer of deoxyribose sugars and phosphates that are joined together via phosphodiester linkages. One of four bases protrudes from each sugar: adenine and guanine, which are purines, and thymine and cytosine, which are pyrimidines. While the sugar phosphate backbone is regular, the order of bases is irregular and this is responsible for the information content of DNA.

Each chain has a 5′ to 3′ polarity, and the two chains of the double helix are oriented in an antiparallel manner—that is, they run in opposite directions. Pairing between the bases holds the chains together. Pairing is mediated by hydrogen bonds and is specific: Adenine on one chain is always paired with thymine on the other chain, whereas guanine is always paired with cytosine. This strict base-pairing reflects the fixed locations of hydrogen atoms in the purine and pyrimidine bases in the forms of those bases found in DNA. Adenine and cytosine almost always exist in the amino as opposed to the imino tautomeric forms, whereas guanine and thymine almost always exist in the keto as opposed to enol forms.

The complementarity between the bases on the two strands gives DNA its self-coding character. The two strands of the double helix fall apart (denature) upon exposure to high temperature, extremes of pH, or any agent that causes the breakage of hydrogen bonds. Upon slow return to normal cellular conditions, the denatured single strands can specifically reassociate to biologically active double helices (renature or anneal). DNA in solution has a helical periodicity of about 10.5 base pairs per turn of the helix. The stacking of base pairs upon each other creates a helix with two grooves. Because the sugars protrude from the bases at an angle of about 120°, the grooves are unequal in size. The edges of each base pair are exposed in the grooves, creating a pattern of hydrogen bond donors and acceptors and of van der Waals surfaces that identifies the base pair.

The wider—or *major*—groove is richer in chemical information than the narrow (*minor*) groove and is more important for recognition by nucleotide sequence-specific binding proteins. Almost all cellular DNAs are extremely long molecules, with only one DNA molecule within a given chromosome. Eukaryotic cells accommodate this extreme length in part by wrapping the DNA around protein particles known as nucleosomes. Most DNA molecules are linear but some DNAs are circles, as is often the case for the chromosomes of prokaryotes and for certain viruses. DNA is flexible. Unless the molecule is topologically constrained, it can freely rotate to accommodate changes in the number of times the two strands twist about each other.

DNA is topologically constrained when it is in the form of a covalently closed circle, or when it is entrained in chromatin. The linking number is an invariant topological property of covalently closed circular DNA. It is the number of times one strand would have to be passed through the other strand in order to separate the two circular strands. The linking number is the sum of two interconvertible geometric properties: twist, which is the number of times the two strands are wrapped around each other; and the writhing number, which is the number of times the long axis of the DNA crosses over itself in space. DNA is relaxed under physiological conditions when it has about 10.5 base pairs per turn and is free of writhe. If the linking number is decreased, then the DNA becomes torsionally stressed, and it is said to be negatively supercoiled.

DNA in cells is usually negatively supercoiled by about 6%. The left-handed wrapping of DNA around nucleosomes introduces negative supercoiling in eukaryotes. In prokaryotes, which lack histones, the enzyme DNA gyrase is responsible for generating negative supercoils. DNA gyrase is a member of the type II family of topoisomerases. These enzymes change the linking number of DNA in steps of two by making a transient break in the double helix and passing a region of duplex DNA through the break. Some type II topoisomerases relax supercoiled DNA, whereas DNA gyrase generates negative supercoils. Type I topoisomerases also relax supercoiled DNAs but do so in steps of one in which one DNA strand is passed through a transient nick in the other strand. RNA differs from DNA in the following ways: its backbone contains ribose rather than 2′-deoxyribose; it contains the pyrimidine uracil in place of thymine; and it usually exists as a single polynucleotide chain, without a complementary chain.

As a consequence of being a single strand, RNA can fold back on itself to form short stretches of double helix between regions that are complementary to each other. RNA allows a greater range of base pairing than does DNA. Thus, as well as A:U and C:G pairing, U can also pair with G. This capacity to form a non-Watson-Crick base pair adds to the propensity of

RNA to form doublehelical segments. Freed of the constraint of forming longrange regular helices, RNA can form complex tertiary structures, which are often based on unconventional interactions between bases and between bases and the sugarphosphate backbone.

Some RNAs act as enzymes—they catalyze chemical reactions in the cell and in vitro. These RNA enzymes are known as ribozymes. Most ribozymes act on phosphorous centers, as in the case of the ribonuclease RNase P. RNase P is composed of protein and RNA, but it is the RNA moiety that is the catalyst. The hammerhead is a self-cleaving RNA, which cuts the RNA backbone via the formation of a 2′, 3′ cyclic phosphate in a reaction that involves an RNA-bound Mg″ ion. Peptidyl transferase is an example of a ribozyme that acts on a carbon centre.

This ribozyme, which is responsible for the formation of the peptide bond, is one of the RNA components of the ribosome. The discovery of RNA enzymes that can act on phosphorous or carbon centers suggests that life might have evolved from a primitive form in which RNA functioned both as the genetic material and as the enzymatic machinery.

Chapter 4

DNA Replication

The process of copying a double-stranded DNA molecule to form two double-stranded molecules DNA replication. The process of DNA replication is a fundamental process used by all living organisms as it is the basis for biological inheritance. As each DNA strand holds the same genetic information, both strands can serve as templates for the reproduction of the opposite strand. The template strand is preserved in its entirety and the new strand is assembled from nucleotides. This process is called "semiconservative replication".

The resulting double-stranded DNA molecules are identical; proofreading and error-checking mechanisms exist to ensure near perfect fidelity. In a cell, DNA replication must happen before cell division can occur. DNA synthesis begins at specific locations in the genome, called "origins", where the two strands of DNA are separated.

RNA primers attach to single stranded DNA and the enzyme DNA polymerase extends the primers to form new strands of DNA, adding nucleotides matched to the template strand. The unwinding of DNA and synthesis of new strands forms a replication fork.

In addition to DNA polymerase, a number of other proteins are associated with the fork and assist in the initiation and continuation of DNA synthesis. DNA replication can also be performed artificially, using the same enzymes used within the cell. DNA polymerases and artificial DNA primers are used to initiate DNA synthesis at known sequences in a template molecule.

The polymerase chain reaction (PCR), a common

laboratory technique, employs artificial synthesis in a cyclic manner to rapidly and specifically amplify a target DNA fragment from a pool of DNA.

ORIGINS OF REPLICATION

The first step in DNA replication is the separation of the two DNA strands that make up the helix that is to be copied. DNA Helicase untwists the helix at locations called replication origins.

The replication origin forms a Y shape, and is called a replication fork. The replication fork moves down the DNA strand, usually from an internal location to the strand's end. The result is that every replication fork has a twin replication fork, moving in the opposite direction from that same internal location to the strand's opposite end. Single-stranded binding proteins (SSB) work with helicase to keep the parental DNA helix unwound.

It works by coating the unwound strands with rigid subunits of SSB that keep the strands from snapping back together in a helix. The SSB subunits coat the single-strands of DNA in a way as not to cover the bases, allowing the DNA to remain available for base-pairing with the newly synthesized daughter strands.

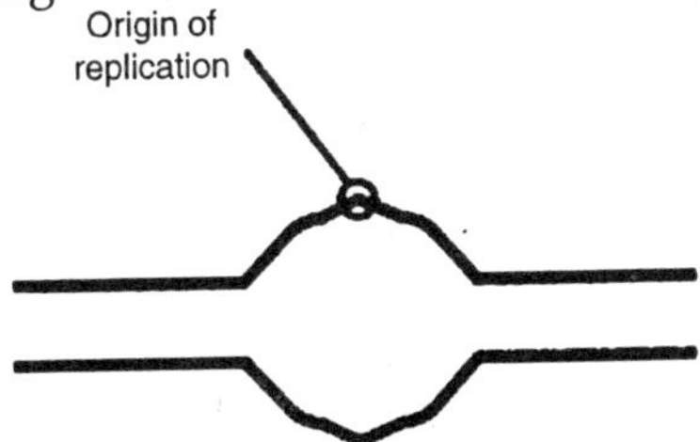

Fig. Replication Fork

As you can see in figure when the two parent strands of DNA are separated to begin replication, one strand is oriented in the 5' to 3' direction while the other strand is oriented in the 3' to 5' direction.

DNA replication, however, is inflexible: the enzyme that carries out the replication, DNA polymerase, only functions in the 5' to 3' direction.

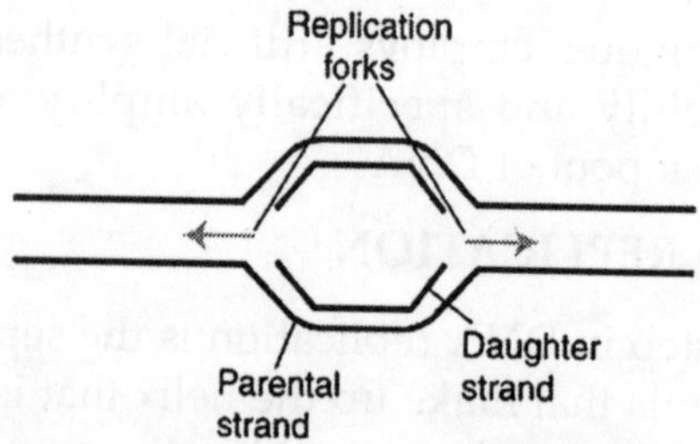

Fig. Replication Fork

This characteristic of DNA polymerase means that the daughter strands synthesize through different methods, one adding nucleotides one by one in the direction of the replication fork, the other able to add nucleotides only in chunks. The first strand, which replicates nucleotides one by one is called the leading strand; the other strand, which replicates in chunks, is called the lagging strand.

DNA REPLICATION IS SEMI-CONSERVATIVE

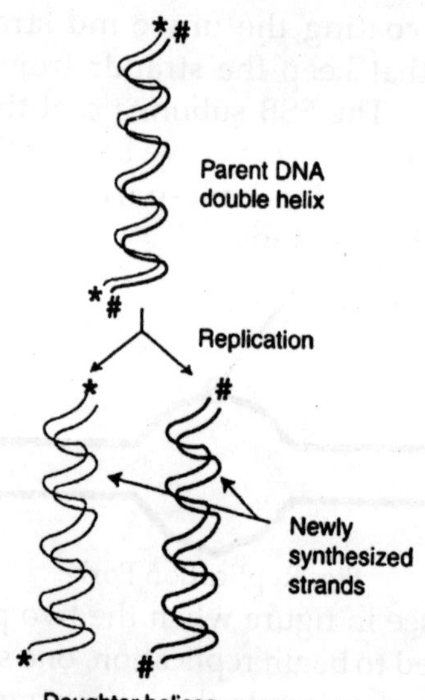

Fig. The Semi-Conservative Nature of DNA Replication

DNA replication of one helix of DNA results in two identical helices. If the original DNA helix is called the "parental" DNA, the two resulting helices can be called "daughter" helices. Each of these two daughter helices is a nearly exact copy of the parental helix (it is not 100% the same due to mutations). DNA creates "daughters" by using the parental strands of DNA as a template or guide.

Each newly synthesized strand of DNA (daughter strand) is made by the addition of a nucleotide that is complementary to the parent strand of DNA. In this way, DNA replication is semi-conservative, meaning that one parent strand is always passed on to the daughter helix of DNA.

THE LAGGING STRAND

Whereas the DNA polymerase on the leading strand can simply follow the replication fork, because DNA polymerase must move in the 5' to 3' direction, on the lagging strand the enzyme must move away from the fork. But if the enzyme moves away from the fork, and the fork is uncovering new DNA that needs to be replicated, then how can the lagging strand be replicated at all? The problem posed by this question is answered through an ingenious method.

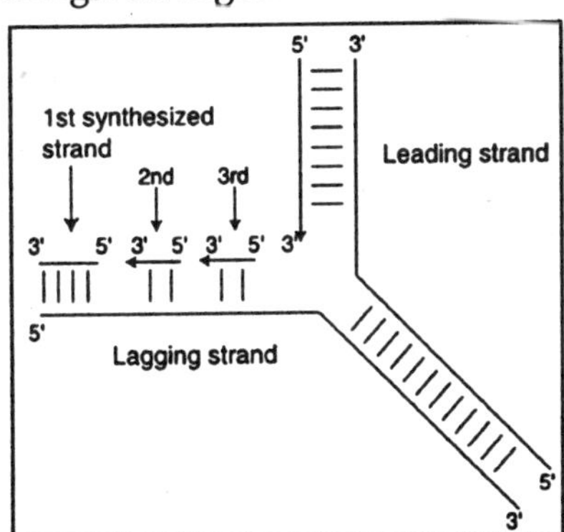

Fig. Leading and Lagging Strands

The lagging strand replicates in small segments, called Okazaki fragments. These fragments are stretches of 100 to 200 nucleotides in humans (1000 to 2000 in bacteria) that are synthesized in the 5' to 3' direction away from the replication fork. Yet while each individual segment is replicated away from the replication fork, each subsequent Okazaki fragment is replicated more closely to the receding replication fork than the fragment before. These fragments are then stitched together by DNA ligase, creating a continuous strand. This type of replication is called discontinuous

As you can see in the figure above, the first synthesized Okazaki fragment on the lagging strand is the furthest away from the replication fork, which is itself receding to the right. Each subsequent Okazaki fragment starts at the replication fork and continues until it meets the previous fragment. The two fragments are then stitched together by DNA ligase.

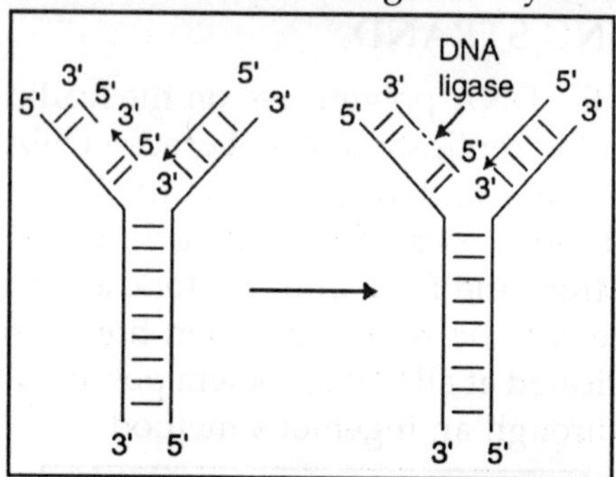

Fig. Patching Up Okazaki Fragments

In figure above, we can also see how replication on the lagging strand remains slightly behind that on the leading strand. Because synthesis on the lagging strand takes place in a "backstitching" mechanism, its replication is slightly delayed in relation to synthesis on the leading strand. The lagging strand must wait for a patch of the parent helix to open up a short distance in front of the newly synthesized strand before it can begin its synthesis back to the end of the daughter strand.

This "Lag" time does not occur in the leading strand because it synthesizes the new strand by following right behind as the helix unwinds at the replication fork. Another complication to replication on the lagging strand is the initiation of replication. Whereas the RNA primer on the leading strand only has to trigger the initiation of the strand once, on the lagging strand each individual Okazaki fragment must be triggered.

On the lagging strand, then, an enzyme called primase that moves with the replication fork synthesizes numerous

RNA primers, each of which triggers the growth of an Okazaki fragment. The RNA primers are eventually removed leaving gaps that are filled by the replication machinery.

THE LEADING STRAND

Since DNA replication moves along the parent strand in the 5' to 3' direction, replication can occur very easily on the leading strand. As seen in figure, the nucleotides are added in the 5' to 3' direction. Triggered by RNA primase, which adds the first nucleotide to the nascent chain, the DNA polymerase simply sits near the replication fork, moving as the fork does, adding nucleotides one after the other, preserving the proper anti-parallel orientation. This sort of replication, since it involves one nucleotide being placed right after another in a series, is called continuous.

DNA STRUCTURE

DNA usually exists in a double-stranded structure, with both strands coiled together to form the characteristic double-helix. Each single strand of DNA is a chain of four types of nucleotide: adenine, cytosine, guanine, and thymine. A nucleotide consists of a phosphate and a deoxyribose sugar forming the backbone of the DNA double helix plus a base that points inwards. Nucleotides are matched between strands through hydrogen bonds to form base pairs.

Adenine pairs with thymine and cytosine pairs with guanine. The physical pairing of bases in DNA means that the information contained within each strand is redundant. The nucleotides on a single strand can be used to reconstruct nucleotides on a newly synthesized partner strand. DNA strands have a directionality, and the different ends of a single strand are called the "3' end" and the "5' end" (these refer to the carbon atom in ribose that the next phosphate in the chain attaches to).

In addition to being complementary, the two strands of DNA are antiparallel: they are orientated in opposite directions. This directionality has consequences in DNA synthesis, because DNA polymerase can only synthesize DNA in one direction by adding nucleotides to the 3' end of a DNA strand.

DNA POLYMERASE

DNA polymerases are a family of enzymes critical for all forms of DNA replication. A DNA polymerase synthesizes a new strand of DNA by extending the 3' end of an existing nucleotide chain, adding new nucleotides matched to the template strand one at a time. Some DNA polymerases may also have some proofreading ability, removing nucleotides from the end of a strand in order to remove any mismatched bases. DNA polymerases are generally extremely accurate, making less than one error for every million nucleotides added. The energy for the process of DNA polymerization comes from the two additional phosphates attached to each of the unincorporated nucleotides.

These free nucleotides, also known as nucleoside triphosphates, contain a total of three phosphates. When a nucleotide is being added to a growing DNA strand, two of the phosphates are removed and the energy produced is used to attach the remaining phosphate to the growing chain. The energetics of this process may also explain the directionality of synthesis - if DNA were synthesized in the 3' to 5' direction, the energy for the process would come from the 5' end of the growing strand rather than from free nucleotides.

During proofreading, if the 5' nucleotide needed to be removed this triphosphate end would be lost, losing the energy source required to add a new nucleotide to the end. DNA polymerase can only extend an existing DNA strand paired with a template strand, it cannot begin the synthesis of a new strand. To do this a short fragment of DNA or RNA, called a primer, must be created and paired with the template strand before DNA polymerase can synthesize new DNA.

DNA REPLICATION WITHIN THE CELL

Origins of Replication

For a cell to divide, it must first replicate its DNA. This

process is initiated at particular points within the DNA, known as "origins", which are targeted by proteins that separate the two strands and initiate DNA synthesis. Origins contain DNA sequences recognized by replication initiator proteins (eg. dnaA in *E coli'* and the Origin Recognition Complex in yeast).

These initiator proteins recruit other proteins to separate the two strands and initiate replication forks. Initiator proteins recruit other proteins to separate the DNA strands at the origin, forming a bubble. Origins tend to be "AT-rich" (rich in adenine and thymine bases) to assist this process because A-T base pairs have two hydrogen bonds (rather than the three formed in a C-G pair)—strands rich in these nucleotides are generally easier to separate. Once strands are separated, RNA primers are created on the template strands and DNA polymerase extends these to create newly synthesized DNA.

As DNA synthesis continues, the original DNA strands continue to unwind on each side of the bubble, forming replication forks. In bacteria, which have a single origin of replication on their circular chromosome, this process eventually creates a "theta structure". In contrast, eukaryotes have longer linear chromosomes and initiate replication at multiple origins within these.

THE REPLICATION FORK

The replication fork is a structure which forms when DNA is being replicated. It is created through the action of helicase, which breaks the hydrogen bonds holding the two DNA strands together. The resulting structure has two branching "prongs", each one made up of a single strand of DNA.

Leading Strand Synthesis

In DNA replication, the leading strand is defined as the new DNA strand at the replication fork that is synthesized in the 5"!3' direction in a continuous manner. When the enzyme helicase unwinds DNA, two single stranded regions of DNA (the "replication fork") form. On the leading strand DNA polymerase III is able to synthesize DNA using the free 3' OH group donated by a single RNA primer and continuous synthesis occurs in the direction in which the replication fork is moving.

Lagging Strand Synthesis

The lagging strand is the DNA strand at the opposite side of the replication fork from the leading strand, running in the 3' to 5' direction. Because DNA polymerase cannot synthesize in the 3"!5' direction, the lagging strand is synthesized in short segments known as Okazaki fragments. Along the lagging strand's template, primase builds RNA primers in short bursts.

DNA polymerases are then able to use the free 3' OH groups on the RNA primers to synthesize DNA in the 5"!3' direction. The RNA fragments are then removed (different mechanisms are used in eukaryotes and prokaryotes) and new deoxyribonucleotides are added to fill the gaps where the RNA was present. DNA ligase then joins the deoxyribonucleotides together, completing the synthesis of the lagging strand.

DYNAMICS AT THE REPLICATION FORK

As helicase unwinds DNA at the replication fork, the DNA ahead is forced to rotate. This process results in a build-up of twists in the DNA ahead. This build-up would form a resistance that would eventually halt the progress of the replication fork. DNA topoisomerases are enzymes that solve these physical problems in the coiling of DNA. Topoisomerase I cuts a single backbone on the DNA, enabling the strands to swivel around each other to remove the build-up of twists. Topoisomerase II cuts both backbones, enabling one double-stranded DNA to pass through another, thereby removing knots and entanglements that can form within and between DNA molecules.

Bare single-stranded DNA has a tendency to fold back upon itself and form secondary structures; these structures can interfere with the movement of DNA polymerase. To prevent this, single-strand binding proteins bind to the DNA until a second strand is synthesized, preventing secondary structure formation.

Clamp proteins form a sliding clamp around DNA,

helping the DNA polymerase maintain contact with its template and thereby assisting with processivity. The inner face of the clamp enables DNA to be threaded through it. Once the polymerase reaches the end of the template or detects double stranded DNA, the sliding clamp undergoes a conformational change which releases the DNA polymerase. Clamp-loading proteins are used to initially load the clamp, recognizing the junction between template and RNA primers.

Eukaryotes

Within eukaryotes, DNA replication is controlled within the context of the cell cycle. As the cell grows and divides, it progresses through stages in the cell cycle; DNA replication occurs during the S phase (Synthesis phase). The progress of the eukaryotic cell through the cycle is controlled by cell cycle checkpoints. Progression through checkpoints is controlled through complex interactions between various proteins, including cyclins and cyclin-dependent kinases.

The G1/S checkpoint (or restriction checkpoint) regulates whether eukaryotic cells enter the process of DNA replication and subsequent division. Cells which do not proceed through this checkpoint are quiescent in the "G0" stage and do not replicate their DNA.

Replication of chloroplast and mitochondrial genomes occurs independent of the cell cycle, through the process of D-loop replication.

Bacteria

Most bacteria do not go through a well-defined cell cycle and instead continuously copy their DNA; during rapid growth this can result in multiple rounds of replication occurring concurrently. Within *E coli*, the most well-characterized bacteria, regulation of DNA replication can be achieved through several mechanisms, including: the hemimethylation and sequestering of the origin sequence, the ratio of ATP to ADP, and the levels of protein DnaA. These all control the process of initiator proteins binding to the origin sequences.

Because *E coli* methylates GATC DNA sequences, DNA synthesis results in hemimethylated sequences. This hemimethylated DNA is recognized by a protein (SeqA) which binds and sequesters the origin sequence; in addition, dnaA (required for initiation of replication) binds less well to hemimethylated DNA. As a result, newly replicated origins are prevented from immediately initiating another round of DNA replication.

ATP builds up when the cell is in a rich medium, triggering DNA replication once the cell has reached a specific size. ATP competes with ADP to bind to DnaA, and the DNA-ATP complex is able to initiate replication. A certain number of DnaA proteins are also required for DNA replication — each time the origin is copied the number of binding sites for DnaA doubles, requiring the synthesis of more DnaA to enable another initiation of replication.

TERMINATION OF REPLICATION

Because bacteria have circular chromosomes, termination of replication occurs when the two replication forks meet each other on the opposite end of the parental chromosome. *E coli* regulate this process through the use of termination sequences which, when bound by the Tus protein, enable only one direction of replication fork to pass through. As a result, the replication forks are constrained to always meet within the termination region of the chromosome.

Eukaryotes initiate DNA replication at multiple points in the chromosome, so replication forks meet and terminate at many points in the chromosome; these are not known to be regulated in any particular manner. Because eukaryotes have linear chromosomes, DNA replication often fails to synthesize to the very end of the chromosomes (telomeres), resulting in telomere shortening.

This is a normal process in somatic cells — cells are only able to divide a certain number of times before the DNA loss prevents further division. (This is known as the Hayflick limit.) Within the germ cell line, which passes DNA to the next generation, the enzyme telomerase extends the repetitive sequences of the telomere region to prevent degradation. Telomerase can become mistakenly active in somatic cells, sometimes leading to cancer formation.

ROLLING CIRCLE REPLICATION

Another method of copying DNA, sometimes used *in vivo* by bacteria and viruses, is the process of rolling circle replication. In this form of replication, a single replication fork progresses around a circular molecule to form multiple linear copies of the DNA sequence. In cells, this process can be used to rapidly synthesize multiple copies of plasmids or viral genomes.

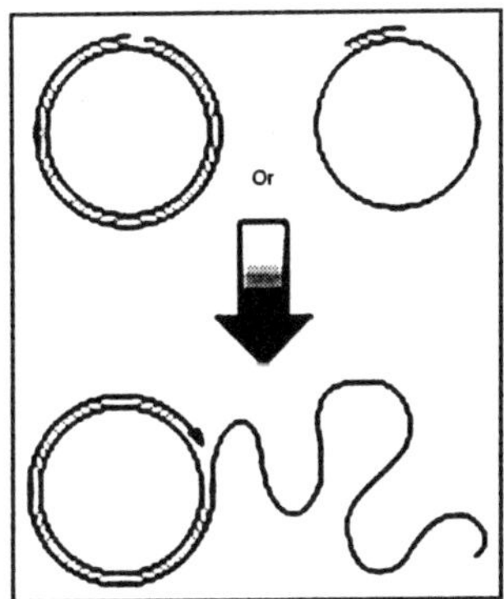

Fig. Rolling Circle Replication

In the cell, rolling circle replication is initiated by an initiator protein encoded by the plasmid or virus DNA. This protein is able to nick one strand of the double-stranded, circular DNA molecule at a site called the double-strand origin (DSO) and remains bound to the 5' phosphate end of the nicked strand. The free 3' hydroxyl end is released and can serve as a primer for DNA synthesis. Using the unnicked strand as a template, replication proceeds around the circular DNA molecule, displacing the nicked strand as single-stranded DNA. Continued DNA synthesis produces multiple single-stranded linear copies of the original DNA in a continuous head-to-tail series. *In vivo* these linear copies are subsequently converted to double-stranded circular molecules.

Rolling circle replication can also be performed *in vitro*

and has found wide uses in academic research and biotechnology, often used for amplification of DNA from very small amounts of starting material. Replication can be initiated by nicking a double-stranded circular DNA molecule or by hybridizing a primer to a single-stranded circle of DNA. The use of a reverse primer (or random primers) produces hyperbranched rolling circle amplification, resulting in exponential rather than linear growth of the DNA molecule.

POLYMERASE CHAIN REACTION

In vitro, researchers commonly replicate DNA using the polymerase chain reaction (PCR). PCR uses a pair of primers to span a target region in template DNA, polymerizing partner strands in each direction. This process can be repeated through multiple cycles through the use of a thermostable polymerase. At the start of each cycle, the mixture of template and primers is heated, separating the newly synthesized molecule and template. Then, as the mixture cools, both of these become templates for new primers to anneal to, and the polymerase extends from these. As a result the number of copies of the target region doubles each round, growing exponentially.

PROKARYOTIC DNA REPLICATION

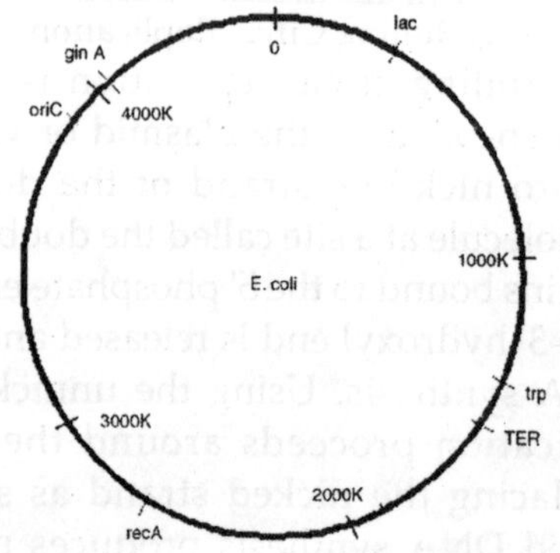

Fig. The Basic Circular Model of the Chromosomes with Map Positions such as ori C and TER Regions

Duplication of DNA is one of the fundamental properties of "molecule of life". Duplication or replication takes place once in a cell cycle. The duration and initiation point differs from one system to another. When conditions are favorable cell cytoplasmic mass increases and when cytoplasmic mass reaches a ratio to that of the cell size to 2L, bacterial cell division is triggered.

To complete its cell cycle it requires hardly 30 -40 minutes. Several genes involved in different steps of cell division have been characterized. Par excellent system to understand molecular process of prokaryotic cell division is E.coli. The size of the E.coli chromosome is $4.6x10^6$ base pairs, but it is compacted by the binding of histone like proteins.

First the nucleoid i.e the chromosome frees from the membrane and unwinds. Such DNA initiates replication and completes in a matter of 15-20 minutes. Then the cytoplasm starts dividing almost in the centre of the cell by central septal ring formation. It is at this time the daughter DNA molecules are partitioned and segregated to the two compartments. The central periseptal ring that started 1-2 minutes after DNA replication now furrows inwards and divides the cytoplasm in the middle region. Then mid-wall splits in the middle and daughter cell are set free, hence the Schizomycetes to bacteria (E.coli)

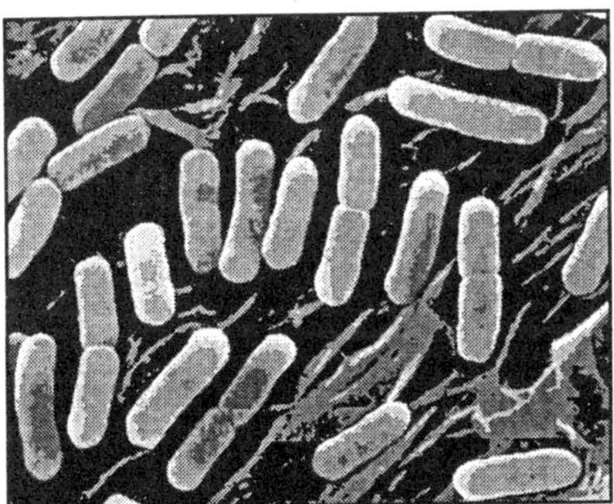

Fig. Bacterial Cells – some Divided and some Yet to be Divided

Bacterial cell division requires the participation of the following genes and gene products. Eukaryotic cell division is more complex and highly regulated. The paradigm for it is yeast, for it is a unicellular system, grows fast and requires hardly 60 minutes for one cell cycle. One can obtain different kinds of mutation like conditional and null or auxotrophic kinds. In culture cells, the duration of one cell cycle is about 12 hrs. Onion root tip cells divide once in 12 hrs.

Whereas embryonic cells in its early stages require just 30 minutes for one cell cycle and they go through 6 to 8 such cell divisions very fast. Replication of DNA in cell division is an important phase. The replication is very accurate and it is nucleotide by nucleotide and there is no provision for making any mistakes, even if mistakes are made they are immediately corrected or repaired, otherwise consequences are serious and deleterious. In general DNA replication is precise, exact and regulated and involves initiation, elongation and termination steps, but mechanisms and rate of replication vary from one system to the other.

REPLICATION IS SEMI CONSERVATIVE

Meselson and Stahl showed that DNA replicates in semi conservative mode. There were some doubts about the mode whether it is conservative, dispersive or semi conservative. But Meselson and Stahl used nonradioactive heavy Nitrogen isotope (15N), as the source of Nitrogen. When cells were grown on such source, 15^N incorporated nitrogen bases get incorporated into DNA, thus it becomes heavier than the DNA that is grown in normal (14N) nitrogen media. When two such DNAs, one grown in the presence of 15N source and the other in 14N (normal), are extracted separately and stained with ethidium Bromide, then subjected to equilibrium density centrifugation at ultra speed of 42000rpm for about 24 –36 hrs, the DNA, separates as two distinct bands, the one that is heavier (15N) is the lower band than the normal (14N) upper DNA.

If cells grown in heavy isotopes are shifted to normal Nitrogen source and grown for one or two cell generation and if DNA is isolated from it and if one combine this with the first two DNA sources and subject them to equilibrium density ultra centrifugation (using cesium chloride or cesium sulphate), one can observe three bands, the top DNA band is from cells grown in 14N without isotope, the second band is the hybrid band with one strand 15^N and the other with normal 14N-Nitrogen and the third band is with heavy nitrogen where both the strands are labeled with 15N.

This experiment virtually and unambiguously proved that DNA replicates in semi conservative mode. Taylor using root tips, labeled with 32^P radioisotopes, of Vicia faba demonstrated semi conservative mode of replication at chromosomal level. Semi conservative means that the two daughter molecules produced at the end of replication, each of the molecules retains one parental strand and the other one is the newly made one. So it is semi conservative. The mode of replication is absolutely semi conservative irrespective the kind of DNA, whether it is ds DNA or ssDNA, whether it is linear or circular, and the basic mechanism is more or less similar; each of them uses their own specialized mechanisms.

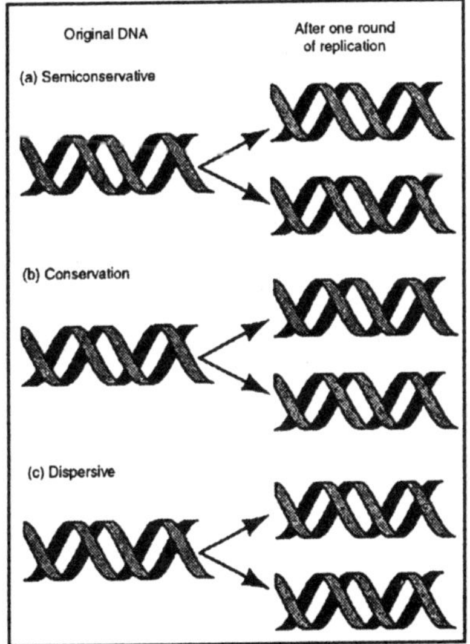

Experimentally semiconservative mode of replication has been shown by density gradient centrifugation of non-radioactive isotopes such as 15N and 14N sources.

DNA REPLICATION

DNA Replication (also known as DNA synthesis) is a process where the double stranded Deoxyribonucleic Acid

(DNA) is copied. Replication is an important first step in cell division, as cells must duplicate their entire genetic constitution before they can divide into two daughter cells. DNA replication is also a critical requirement for DNA repair.

REQUIREMENTS OF DNA REPLICATION

Replication requires three important components in order for this process to work.

Template Strand

The DNA serves as a template to guide the incoming nucleotides. The template strand attaches to the RNA primer strand in order to replicate the strand of DNA. The template just guides the nucleotides to the place that they need to go in. This is another component that is needed in order for DNA replication to even take place. If the DNA only had the template strand and none of the rest of the things then the strand could not replicate. The template starnd is very thin. Through an enzyme called primase the DNA template strand is synthesized one nucleotide at a time.

DNA Polymerase

There are many DNA polymerases within a cell, but only one of which is used in the replication of DNA. In order for DNA to replicate, it must have a primer, which is a small strand of RNA. The DNA template strand synthesizes one nucleotide at a time by the use of primase. The adding of nucleotides to the 3′ end, which is continued until that section of DNA is replicated. The other DNA polymerases that are within the cell are used in other ways besides replication. DNA polymerase III is one certain polymerase, which aids in replication of what is known as the "lagging strand". DNA ligase is the final enzyme, which combines the lagging strand

Free 3' Hydroxyl

The free 3' hydroxyl is the starter strand called a primer which is required in order for DNA to be replicated. The primer strand is not very big, it is shorter then a single strand of RNA.

The primer is the complementary strand to the template strand of DNA. This is another one of the things that are needed in order to make the process of DNA replication even possible. The DNA polymerase is added to the 3' end of the primer, which keeps growing until the section is finally complete. After a while the RNA primer is degraded and disappears.

Types of Replication

During the discovery of DNA replication they also discovered that there are three modes of replication that a DNA strand can take. The three types of DNA replication are semiconservative, conservative, and dispersive. These three types of replication were tested on in order to see the possible patterns that would result in the complementary base pairing.

Semiconservative Replication

Through this type of replication the parent strand serves as a template for the new strand. The two offsprings would have one of the parent strand and one new strand.

Conservative Replication

Through this type of replication the double helix serves as a template. This although does not contribute to the new double helix.

Dispersive Replication

In this type of replication the fragments from the parent DNA molecule. The DNA serves as a template for the assembly of the new molecule. The new double helix contains old and new parts of the DNA strands.

DNA Repairing

After the DNA is replicated, there are three DNA repair mechanisms. They are to be used when they feel necessary. This will lessen the chance of human mutation. The rate of error in which the DNA polymerases makes a mistake is very high. This is why the DNA has mechanisms to cause fewer mutations. The three mechanisms are:

Proofreading

Every time a nucleotide is introduced to an already growing chain, the DNA polymerase checks the connection. If the pair is mismatched, it will be removed and redone. This lowers the overall rate of mutations.

Mismatch Pair

After the replication takes place, another set of proteins check to make sure that there are no more mismatched base pairs and also which strand is the wrong one.

Excision Pair

There can still be damage while the cell is living. Because of this, there are enzymes constantly checking the cell. If they detect a problem, the enzyme cuts the strand and also cuts away the opposite base. DNA polymerase and ligase then fix this base sequence.

LABORATORY TECHNIQUE

Scientists use two techniques in a laboratory that involve DNA replication. They use these techniques to observe genes and genomes. One of the techniques can make multiple copies of DNA from only a short piece. Another technique allows scientists to determine the base sequence of DNA. The two techniques are:

Polymerase Chain Reaction (PCR)

Through this process, a short strand of DNA is copied repetitively. There are three steps that are repeated in the process. The DNA strands are denatured or heated in order to separate the strand. Next, they add a primer, which was artificially made. They also add DNA polymerase and the four deoxyribonucleotide triphosphates. These three are needed in order to replicate DNA. The final step is then the DNA polymerase catalyzes the production of a complementary strand.

DNA Sequencing

This allows scientists to determines the base sequence of

a DNA. The process is the DNA is first denatured and a single strand is placed in a test tube. They then add DNA polymerase, primer, the four dNTP's, and small amounts of ddNTP's. The DNA replicates within the tube and after the DNA fragments are denatured. The fragments then go through electrophoresis, which sorts the lengths of DNA fragments. The fragments then pass through a laser beam which excites the fluorescent tags. This information is fed into a computer at which tells the sequence.

Mendelson- Stahl Experiment

Through this experiment Meselson- Stahl convinced the scientists that the correct model for DNA replication was the semiconservative replication. They used density labeling in order to distinguish the old and new strands of DNA. In their experiment they used "heavy" isotopes of nitrogen. This nitrogen is a nonradioactive isotope that made the molecule more dense. The chemically identical molecules containing an isotope known as ^{14} N. they needed a way to distinguish the different DNA densities.

Because of this they needed to come up with something to measure the densities of solutions. Meselson, Stahl, and Jerome Vanguard came up with centrifuge, which is a procedure that involves the spinning of solutions at high speed, which causes the particles to separate and form gradient according to the different densities of the solution. The first part of their experiment they grew cultures of Escherichia coli. One of the cultures was grown in ^{15} N, which made the DNA "heavy".

The other culture was placed in a medium using ^{14}N rather than ^{15}N, which made all the DNA "light". In order to see if the use of centrifuge worked they combined the two cultures and when centrifuged it formed two DNA bands. This showed them the different densities with in the two cultures. The next part of their experiment they grew E. Coli in the ^{15}N medium and then transferred the bacteria from that medium to the ^{14}N. Through this they came up with the conclusion that through E. Coli DNA replicates every 20 minutes. They then took DNA samples from each generation through, which they found that the density gradient was different each generation.

Their observations could only be explained with the semiconservative model of DNA replication. Their results showed that when the DNNA first underwent replication, it was in 15N, which caused the DNA to be heavy. Because in the semiconservative model of DNA replication, the one strand acts as a template for a second strand which the DNA was in a ^{14}N DNA strand and a ^{15}N and they were of intermediate density.

Chapter 5

Enzymology of DNA Replication

One group of enzymes involved in DNA replication which have been studied in great detail are the DNA dependent DNA polymerases. At least three are known which catalyse the synthesis of a complementary structure using DNA as a template. The first of these. DNA polymerase I, was isolated and characterized by Arthur Kornberg for which he was awarded the Nobel Prize in 1959. Till the early sixties, it was believed that this was the only enzyme with the ability to catalyse DNA synthesis.

Later in the sixties, Roy Curtiss and others showed that bacterial mutants lacking this enzyme still had the ability to replicate DNA and since then, two other DNA polymerases (DNA polymerase II and DNA polymerase III) which can copy DNA have been recognized. In addition to these two enzymes, it is now known that a ligase (joining enzyme), and a primase (to prime the synthesis), in addition to a topoisomerase and other replication proteins are also involved.

DNA polymerase I, was thought to be the major enzyme that joins together de-oxynucleotides in vivo but now it is realized that DNA polymerase III is the main polymerizing enzyme while DNA polymerase I is a repair enzyme and fills In the gaps between the small fragments. The role of DNA polymerase II is not yet clear. All these three known polymerases extend the DNA molecules from the 5 end to the 3 end only. This led to the idea that the replication of double stranded DNA occurs in short pieces discontinuously and these short segments are later joined with the aid of a ligase to form a continuous strand.

Also, the DNA polymerases can add nucleotides only to a perfectly base paired nucleotide sequence and they cannot initiate new DNA synthesis. It now appears that the primary reaction is carried out by a DNA dependant RNA polymerase, which synthesizes short degradable RNA primers. It is also believed that in addition to the RNA polymerases, specific "primases", which differ from the RNA polymerases, synthesize the primary RNA which is then elongated by the DNA polymerase such a primase has been identified in the bacterium E. coli and also in the bacteriophages T4 and T7 Because of the difficulties in understanding the complex process of replication of bacterial double stranded DNA, investigators have used the single stranded DNA bacteriophage Øx174 as a test system to understand the exact mechanism of DNA replication.

In this virus, a complementary strand (-) is first synthesised on the parental single strand (+ strand) to yield a double helical structure. The new strand (- strand) then serves as the template for the synthesis of more of + strands which are then incorporated into the viral particles. The process appears to be straight forward but is not as simple as stated. Discontinuous synthesis appears to occur even during this type of replication.

Evidence for discontinuous replication of DNA in bacteria first came from Okazaki in 1967, who was able to isolate short pieces of' DNA during replication. It now appears that not all replicating DNA molecules rely on RNA priming and Okazaki fragments to solve the problem posed by the properties of DNA polymerase. The involvement of t-RNA primers in the synthesis of DNA copies of tumor viruses has also been reported. Recently an explanation to describe the mechanism by which single stranded DNA in certain animal viruses replicates has been given.

It is found that in single stranded DNA, "inverted repeats" (hair pin like structures) at each end of the molecule act as primers for initiation of replication. The secondary structure of the DNA is a double stranded helix. For replication to occur semi conservatively, the structure must undergo unwinding.

Over the years, a number of DNA binding proteins, (DNA destabilizing proteins and untwisting enzymes) have been found both in procaryotic and eucaryotic cells and are in-volved in opening up of the double stranded structure to allow replication.

The latest addition to the list are the topoisomerases including the gyrase first isolated from E. coli. This new enzyme catalyses the introduction of negative super coils into double helical DNA in an ATP dependant reaction and appears to be essential for in vivo replication of DNA both in bacteria and bacteriophages. Antibiotics such as nalidixic acid and novobiocin inhibit this enzyme and there by prevent DNA replication at or beyond the replication point.

It is believed to relieve the positive supercoiling strains which builds up during replication and aid in unwinding of the double helix. One other point of interest yet to be solved with regard to DNA replication is the origin of replication and the number of replicating points per DNA molecule. Although a number of origins of replica-tion have been sequenced the mechanism of control is not yet clear. It is however, generally agreed that a DNA molecule may have more than one replicating points from where replication can be initiated.

The base sequence at the origin is not identical but has similarities. A variety of models have been described in literature to explain the mechanism of vivo DNA replication. Unfortunately, none of these provide a complete answer to the problem. The one that has received much attention an d is close to acceptance is the rolling circle model. According to this model, replication starts with a specific cut in one strand of the parental duplex molecule.

This generates a terminal nucleotide with a free 3' OH group while the other end has a phosphate group at the 5' end. As replication proceeds, the 5' end of the open strand is rolled out as a free tail of increasing length. The replicating structure is called a rolling circle since the unraveling of the free single strand is accompanied by a rotation of the double helical template about its axis. The 5' end tail serves as a template for the synthesis of small DNA fragments which are eventually joined together by the DNA ligase. Such growing tails have a double stranded character soon after their formation. Elongation of such tails sometimes goes on to produce tails many times the length of the original circle. It is believed that the tails are cut by specific endonucleases and this is followed by circularization by pairing between the sticky ends.

One of the prevailing fashions in bioscience these days is the application of genomics to eukaryotic gene expression. Largely eclipsed are the approaches of a few decades ago in which enzymes derived from microorganisms blazed the trail to much of our current understanding of macromolecular biosynthesis and gene regulation. In this Commentary I will give an anecdotal account of the lessons I learned from my attempts to resolve and reconstitute biological events in DNA replication and reflect on how these lessons may still apply to solving the current problems of growth and development and the aberrations of disease.

The beginning of the 20th century saw the birth of modern biochemistry with the demonstration that alcoholic fermentation could be observed in the juice of yeast cells. This led to the discovery of the dozen enzymes that convert sucrose to alcohol and ultimately to the reconstitution of alcoholic fermentation at a refined molecular level. Along with these early biochemical studies with yeast came the discovery that virtually the same enzymes and pathway were responsible in mammalian cells for the conversion of glycogen to lactic acid, which provides ATP energy for muscle contraction.

This astonishing fact, along with many other such examples in metabolic and biosynthetic pathways, made it clear that mechanisms and molecules have been preserved in bacteria, fungi, plants, and animals, essentially intact through billions of years of Darwinian evolution. I regard this insight as one of the great revelations of the 20th century.

RELY ON ENZYMOLOGY TO CLARIFY BIOLOGIC QUESTIONS

Based on the conviction that all reactions in the cell are catalyzed and directed by enzymes, the first commandment commands that enzymology can be relied on to clarify a biologic question. Chemists once bridled at this. But time and again, spontaneous reactions, such as the melting of DNA and the folding of proteins, are found to be driven and directed by enzymes; in the case of DNA, its melting in a cell is catalyzed by several different helicases. The first and crucial step is to find a way to observe the phenomenon of interest in a cell-free system.

Should that succeed, then one should be able to reduce the event to its molecular components by enzyme fractionation. This confidence is derived from the fact that, as mentioned, alcoholic fermentation, which had eluded understanding for centuries, was clarified by fractionation of a cell-free yeast extract, as was glycolysis by fractionation of muscle extracts and in the same vein luminescence in the extracts of a firefly and replication of DNA in microbial cell lysates.

Fractionation procedures for these extracts revealed the molecular mechanisms and machines for the catalysis and regulation of many complex reactions and pathways, as recounted here for DNA replication.

With a cell-free system in hand that recreates a biologic event, the biochemist should be able to perform the process as well as the cell does it. Even better! After all, the cell is under great constraints to provide a consensus medium that supports thousands of diverse reactions, only some of which operate under optimal conditions. By contrast, the biochemist enjoys the freedom to saturate each enzyme with its substrate, trap the products, and provide the optimal pH and salt and metal ion concentrations.

The biochemist can thus be creative and effective in analyzing the molecular basis of a reaction or pathway. The refinement of methods to purify proteins by chromatography and to establish their homogeneity by gel electrophoresis, when combined with the power of reverse genetics and genomics, has made the isolation and large-scale preparation of enzymes relatively easy compared to what it was years ago.

Despite this, discrete events in the proliferation,

differentiation, and adaptations of cells and organisms are almost always analyzed by genetic means.

Striking phenotypes are produced by mutations and transfections, but the alterations in enzymes and pathways are generally only inferred. Rarely are they verified by the isolation of proteins with demonstrable functions. To many cell biologists and developmental biologists, the need to examine an event in a cell-free system does not come up on their radar screen.

TRUST THE UNIVERSALITY OF BIOCHEMISTRY AND THE POWER OF MICROBIOLOGY

The universality of biochemistry from microbes to humans in basic metabolic and biosynthetic pathways has led to the silly quip: "What's true for *E. coli* is true for elephants, and what's not true for *E. coli* is not true." My faith in this universality encouraged me to focus on how prokaryotes, particularly *Escherichia coli,* replicate their own genomes and those of their phages and plasmids. Microbial generation times, unlike those of eukaryotes, are measured in minutes rather than hours and days. This is where the light on replication shines brightest. I made the choice to work with prokaryotes with the confidence that these systems would be reliable prototypes for how the so-called higher organisms replicate their DNA.

In recent years, exciting advances have been made in discovering and characterizing the eukaryotic replication enzymes: the helicases, topoisomerases, polymerases, primases, ligases, and other components of the chromosomal replicases.

The variations from prokaryotic enzymes are fascinating. Yet virtually all these enzymes and mechanisms were already familiar and adhere to the basic themes discovered earlier in the prokaryotic systems.

DO NOT BELIEVE SOMETHING BECAUSE YOU CAN EXPLAIN IT

In 1950, having found the enzymes that incorporate

nucleotides into coenzymes and curious about how they might become part of nucleic acids, I needed first to determine how the purine and pyrimidine bases became substrates for assembly into these polymers. In the course of exploring the biosynthesis of nucleotides, I learned how to use labeled bases and how to tag each of the phosphates of the nucleoside diphosphates (NDPs) and triphosphates (NTPs). In 1954, we observed an activity in an extract of *E. coli* that incorporated the label of [á-^{32}P]ATP into an acid-insoluble form that we presumed to be RNA.

While making progress in purifying this activity, we learned of a discovery in the laboratory of Severo Ochoa in New York. While observing an exchange of orthophosphate with ADP in extracts of *Azotobacter vinelandii*, they discovered an enzyme that converted the ADP and other NDPs into an RNA-like polymer.

Acting on this information, we substituted ADP for ATP and found that our activity was far greater. Clearly ADP was the preferred substrate over ATP. The *E. coli* enzyme we then purified was the same polynucleotide phosphorylase that the Ochoa laboratory had first identified in *Azotobacter*. As was learned later, the role of the phosphorylase was to degrade RNA rather than effect its synthesis. Had we persisted with ATP as substrate, we would surely have found RNA polymerase, the true synthetic enzyme, a year earlier than we did DNA polymerase and several years before it was discovered in 1961 by the late Sam Weiss.

CLEAN THINKING ON DIRTY ENZYMES

The late Efraim Racker enunciated this commandment, and I have been one of its ardent disciples. A dramatic example is the discovery of DNA replication. I first observed DNA synthesis in an *E. coli* extract in 1955, when I found that 50 counts out of a million of thymidine were incorporated into an acid-insoluble form. Those few counts above background seemed real because they were susceptible to DNase.

Could we possibly figure out what was going on in so crude a system, let alone in an intact cell? We identified and purified the first DNA polymerase, but not before our fractionation procedures disclosed a variety of novel enzymes that acted initially on the DNA, the [^{14}C]thymidine, and the ATP in our incubation mixtures. First, the [^{14}C]thymidine we added had to be phosphorylated by ATP via a new enzyme, thymidine kinase, to become thymidylate.

The added calf thymus DNA proved to be a substrate for DNases that produced the four hitherto unknown deoxynucleoside 52 -monophosphates. These were then phosphorylated by four distinct nucleotide kinases to the corresponding diphosphates, which were in turn phosphorylated by nucleoside diphosphate kinase to the respective and previously unknown dNTPs.

The DNA we had added served three additional functions beyond being a source of the four building blocks. It was a template to direct the precise order of nucleotide assembly, a source of primer termini for chain elongation, and a pool to protect the tiny amount of synthesized DNA from degradation by the nucleases that are abundant in cell extracts. The *E. coli* extract was thus the source of seven new enzymes in addition to the enzyme we named DNA polymerase.

With that, template and primer were introduced into the language of all polymerase actions. We learned all this from fractionating the *E. coli* extract into its multiple activities and finally putting the purified enzymes and their products back together. These studies taught us the basic features of how DNA polymerases act and, incidentally, that the strands of duplex DNA are oriented in opposite directions, not known at the time. Cell extracts are by their nature "dirty enzymes"; intact cells and organisms are "dirtier" still. F. G. Hopkins, a prescient pioneer in the biochemical basis of nutrition, said it best back in 1931: "(The biochemist's word) may not be the last in the description of life, but without his help the last word will never be said."

And so it has been with many cellular events, most recently the fully reconstituted transcription by the 48-subunit yeast RNA polymerase II initiation complex and the sorting of proteins to specific subcellular compartments by fractionated vesicles and enzymes. Purification of an enzyme to homogeneity now opens the door to reverse genetics, and the enzymes themselves still provide unique reagents, as commandments IX and X will describe. Sometimes, an apparently pure enzyme may be found on further purification to harbor a contaminant of great importance.

As one example, an extra step in the purification of DNA polymerase I rendered the enzyme inactive. The reason: the template-primer used in our routine assays was DNA activated by being nicked many times by a DNase. We were not aware at the time that these nicks had been enlarged upon by an exonuclease in our polymerase preparation to create a stretch of exposed template needed by the polymerase. The exonuclease activity we had fractionated away, which we then purified and named exonuclease III, proved to be a crucial reagent in the discovery of recombinant DNA.

In 1967, after 10 years of trying and failing to prove that our enzymatically synthesized DNA was biologically active, we finally succeeded. What made the difference was the use of a single-stranded, circular DNA of a bacterial virus as template and the discovery of DNA ligase that could circularize the linear product. With X174 DNA, we could make quantities of the circularly closed, infectious viral DNA. The sequence of 5,386 nucleotides was correct, and there was no need for novel nucleotides or other components, as had been conjectured.

We also pointed out that this in vitro system afforded the means to introduce novel nucleotides for site-directed mutagenesis. The appearance of our paper announcing the test tube synthesis of infectious DNA generated a huge crush of media attention, a congratulatory phone call from President Lyndon Johnson and headlines worldwide, all based on the belief that we had synthesized a big, hairy virus and "created life in the test tube." I had to explain to the assembled reporters that it was not I who assembled the long DNA chain of the virus but rather it was the awesome enzyme, DNA polymerase, that I had identified and isolated from *E. coli* cells.

And further, I had to make it clear that it was these

bacteria in a culture flask that imbibed the viral DNA to make the infectious virus particles. As for "creation of life in the test tube," some might dispute that a virus is even a "living" creature.

DO NOT WASTE CLEAN ENZYMES ON DIRTY SUBSTRATES

Three years after the hoopla about the synthesis of a viral DNA by our DNA polymerase, serious questions remained. How is a DNA chain started? How is the accumulating genetic evidence for additional polymerases and other factors needed for replication explained? A cartoon at the time showed the apparatus at a replication fork discreetly obscured by a fig leaf, and there were polemical attacks in *Nature New Biology* that dismissed our DNA polymerase as merely a repair enzyme with little relevance to replication, in essence a "red herring." Over the years, we had tried a variety of DNA samples to demonstrate the start of a DNA chain.

The results were negative or equivocal. Then it dawned on me that we were violating the fifth commandment. We were using a pure DNA polymerase on a dirty DNA substrate: frayed, gapped, fragmented, denatured and heterogeneous. When we finally switched to the intact, single-stranded, circular DNA of a small bacteriophage, we discovered how a DNA chain is started: priming with RNA.

Single-stranded phages provided not only the DNA substrate with which we could discover the RNA priming of new chains, but also the enzyme systems responsible for the priming and subsequent replication. The filamentous phage M13 depends on the host RNA polymerase to make a short transcript of an origin region, whereas the icosahedral phage X174 appropriates a complex primosome, used by the host to prime the start of chains on the lagging strand at the replication fork.

Whereas the conversion of the single M13 viral strand to the duplex replicative form was readily resolved and reconstituted, the conversion of the X174 single-stranded circle was far more complex and required the discovery of 15 new proteins, which constitute the apparatus at the host chromosomal replicating fork. With these many proteins in hand we could attempt to discover how replication was initiated at the origin of the intact *E. coli* chromosome.

CORRECT FOR EXTRACT DILUTION WITH MOLECULAR CROWDING

Cells are gels. Half of the cell dry weight is made up of proteins packed in highly organized communities. That some of their functions, individually and collectively, can be observed despite great dilution (20-fold or more) is a fortunate break for biochemistry. But there is an absolute need in some cases to restore the crowded molecular state, as we learned from our attempts to observe initiation of replication at the origin of an intact chromosome. We were given a 5-kb plasmid containing the origin of the 4,000-kb *E. coli* chromosome that is replicated in the cell with the physiological and genetic features of the host chromosome, in effect a minichromosome.

When Seichi Yasuda came from Japan with this *oriC* plasmid, I thought we would soon resolve and reconstitute its replication much as we had done with phage X174. But it took 10 man-years of utter frustration before we finally succeeded in making a cell-free system work. Success in achieving *oriC* plasmid replication in a cell-free state depended on two strange maneuvers. One was to include a high concentration of polyethylene glycol (PEG) (10% [wt/vol]), 10,000 Da) in the incubation mixture. As is true of such hydrophilic polymers, the PEG gel occupies most of the aqueous volume and excludes a small volume into which large molecules are crowded.

This concentration is essential when several proteins are needed in the consecutive steps of a pathway. The other maneuver repeated an earlier experience in which we proceeded to fractionate an inactive lysate with ammonium sulfate. Progressive additions of the salt yielded precipitates, in one of which the active proteins were present and concentrated when dissolved in a small volume. Just as important, the supernatant fraction we discarded contained a potent inhibitor, a nuclease that relaxed the plasmid DNA from its essential supercoiled state. Along with a purified, origin-binding DnaA protein, we provided primosomal, replication, and ligase proteins with other factors to obtain rapid, origin-specific, extensive replication of the *oriC* minichromosome. With these many proteins added in sufficient amounts, PEG was no longer needed. The mechanisms we discovered in *oriC* replication were found to apply to the replication of many microbial plasmids and phages and to some eukaryotic viruses and episomes.

RESPECT THE PERSONALITY OF DNA

For years, DNA was regarded as a rigid rod devoid of personality and plasticity. Only upon heating did DNA change shape, melting into a random coil of its single strands. Then we came to realise that the shape of DNA is dynamic in ways essential for its multiple functions.

Chromosome organization, replication, transcription, recombination, and repair have revealed that DNA can bend, twist, and writhe, can be knotted, catenated, and supercoiled (positive and negative), can be in A, B, and Z helical forms, and can breathe. Especially noteworthy is breathing, the transient thermodynamic-driven opening (melting) of the duplex that facilitates the binding of specific proteins such as the helicase responsible for priming and the onset of replication.

Certain DNA sequences are also predisposed to a more extensive form of melting ("heavy breathing") that creates a relatively large opening for transcription. The resulting RNA-DNA duplex (R-loop) can activate an inert origin of replication by altering its structure, even hundreds of base pairs away, which facilitates its opening by origin-binding and replication proteins. Negative supercoiling supplies the energy for the breathing and other features that direct the shape and movements of DNA at the *oriC* origin of replication. These DNA responses have led to an appreciation of the role of transcriptional activation of replication origins near primers in large chromosomes, both prokaryotic and eukaryotic.

USE REVERSE GENETICS AND GENOMICS

Direct genetics, in which a randomly mutated gene can ultimately be linked to a deficiency in a single enzyme, was a landmark discovery in biologic science. This approach served well by providing *E. coli* mutants defective in replication, some in initiation of a chromosome (e.g., *dnaA*) and others in elongation (e.g., *dnaB, dnaC, dnaE,* and *dnaG*). But randomly generated mutants do not readily disclose the products of their genes nor their particular functions.

Nevertheless, these replication mutants were crucial in validating our assays because DNA synthesis was absent in the extracts of mutant cells and restored when extracts or purified fractions from wild-type cells were added. Reverse genetics and genomics have now made the enzymologic approach even more powerful. Unlike direct genetics, enzymology starts with a defined function, after which finding the responsible genes has become relatively easy.

With even a picomole of a purified enzyme or a band on a gel, a peptide sequence can be determined and the encoding gene identified, cloned, and overexpressed; genomics facilitates the process by providing the complete genome sequences of *E. coli*, yeast, and many other microbes. Profound insights into the physiologic role of an enzyme or pathway emerge from the behaviour of cells with a null, point, or truncated mutation of a gene or modulated levels of its overexpression. The ease with which large quantities of pure enzymes can be produced by overexpression of a cloned gene has made their use as reagents even more attractive (commandment X).

"DNA shuffling", a new technique, has made enzyme reagents compelling. By creating a very large number of random rearrangements of a gene or genome, a particular gene product can be selected for a desired property (e.g., heat resistance) with wide applications in industry and biomedical science.

ENZYMES AS UNIQUE REAGENTS

Biochemistry is replete with examples in which enzymes

have been employed as analytic and preparative reagents. From basic research to industrial processes, proteases, amylases, phospholipases, kinases, and phosphatases, etc., have been crucial in operations that were beyond the capabilities of available chemical technology.

I will mention just a few examples of applications to DNA and its replication and one from my recent research on inorganic polyphosphate (poly P). The key discovery in 1944 that identified DNA as the genetic substance was based on the destruction by crystalline pancreatic DNase of the factor that transformed one strain of *Pneumococcus* sp. to another.

It was the action of this DNase again, as mentioned in commandment IV, which in 1955 made me believe that the few counts of [^{14}C]thymidine incorporated by an *E. coli* extract into an acid-insoluble form signaled the synthesis of DNA. Many more examples can be cited in which an enzyme reagent was decisive: the circularization of linear DNAs by ligases, the creation of "sticky" tails by specific exonucleases used to prepare the first recombinant DNAs, the innumerable uses of restriction nucleases, and on and on.

Enzyme reagents have been decisive in my approach to determine the functions of poly P, an inorganic polymer of hundreds of phosphate residues linked by "high-energy" anhydride bonds. Likely present on prebiotic earth, poly P is now found in every living cell, but for lack of any known functions, was earlier regarded as a "molecular fossil."

True to the first commandment, I have sought and isolated enzymes that make and act upon poly P. With these enzymes we developed assays that are definitive, facile, and sensitive in place of those that are ambiguous, laborious, and insensitive. Together with the use of reverse genetics, we have learned that many microbes need poly P to adapt to adverse conditions and to survive in the stationary phase.

The kinase that makes poly P from ATP is highly conserved in some of the major pathogenic bacteria, and mutants lacking the kinase are defective in motility and virulence. Thus, this enzyme, absent from eukaryotes, may prove to be an attractive target for antimicrobial drugs.

ENZYMOLOGY OF THE REPAIR OF FREE RADICALS-INDUCED DNA DAMAGE

Two main pathways appeared during evolution: the release of oxygen by photosynthesis and the aerobic respiration. These highly efficient metabolic systems produce useful, although dangerous molecules for the cells: free radicals (NO· ¼) and reactive oxygen species (ROS) such as $O_2{\cdot}^-$, H_2O_2 and OH·. These molecules are constantly formed in cells by the cellular metabolism and by spontaneous chemical degradation of some biomolecules.

Free radicals are implicated in a wide range of cellular processes but an excess of such molecules could have deleterious effects on living cells leading to cell injury or cell death. Oxidative damage of main cellular components such as DNA, lipids, carbohydrates and proteins has been implemented in the ageing phenomena, ischemia, cancer, autoimmune diseases and neural cell death.

Activation of oncogenes like c-*myc* and ras can also induce ROS by alteration of specific metabolic pathways. Organisms are equally exposed to exogenous factors such as ionising radiation and chemical cancerogens, which can also generate ROS. Evidence has accumulated that lack of protection against free radicals and lack of repair of oxidative damage in biological macromolecules have a significant role in mutagenesis and consequently on carcinogenesis.

Thus maintenance of the cellular equilibrium between prooxidant (ROS) and antioxidant species should be tightly regulated in cells. Specialised systems have evolved, involving cellular antioxidants and DNA repair, to protect cells against ROS-induced injury. Indeed, the defence system, such as DNA repair, is highly conserved from bacteria to human. It is generally assumed that most oxidative DNA damages - base damage, sugar damage and abasic sites - are dealt with by BER. A number of recent reviews report the main features of the BER pathway.

The goal of DNA glycosylases is to locate fast and efficiently the aberrant base amongst a huge excess of normal ones. Very little is known about how these proteins achieve this goal. The comparison of the crystal structures of a number of DNA glycosylases revealed structural homologies leading to the concept of a superfamily of BER DNA glycosylases, the helix-hairpin-helix (HhH) superfamily, having similar HhH fold and a Gly/Pro-rich stretch with nearby Asp (GPD) motifs, although very little sequence similarity. This HhH motif plays an important role in the flipping out of the modified base.

The rate of repair measured for the excision of modified bases is not always optimal and should be improved by the identification and the use of accessory proteins. The recent identification of new DNA polymerases able to replicate efficiently and accurately miscoding and modified bases have to be taken into account in the understanding of BER.

FREE RADICAL SPECIES AND OXIDATIVE DAMAGE OF DNA

DNA has a limited chemical stability (intrinsic or induced by exogenous agents) and is one of the most biologically important targets of ROS. Maintenance of its integrity is a major goal for cells. About 100 different kinds of base and sugar damage have been identified. Free-radicals can damage nucleobases and sugar units in DNA either directly, or indirectly.

Hydroxyl radicals, which are the most active species, predominantly, react with the C_8 of purines forming 7,8-dihydro-8-oxo-2'-deoxyguanosine (8-oxoG) and imidazol ring-opened products such as 2,6-diamino-4-hydroxy-5-formamidopyrimidine (Fapy); with C_5-C_6 double bond of pyrimidines forming glycol and pyrimidines hydrates and with C8-C5' of purines forming 8,5'-cyclopurine deoxynucleosides. The abstraction of a hydrogen atom from deoxyribose at C1' and C4' generates DNA-strand breaks with 3'-phosphoglycolate ester and 3'-phosphate.

Indirectly, ROS can generate reactive aldehydes, as a product of membrane lipids peroxidation, which react with DNA bases forming the exocyclic adducts 1,N^6-ethenoadenine, 1,N^2-ethenoguanine, N^2,3-ethenoguanine, and 3,N^4-ethenocytosine and pyrimidopurinone such as M_1G. Damaged bases such as 8-oxoguanine; 5-hydroxy-2'-deoxycitidine, hypoxanthine, ethenoadducts and pyrimido-purinone have miscoding properties and, if not repaired, lead to mutation upon replication.

Others such as oxidised deoxyribose, formamidopyrimidine, fragmented thymine and thymine glycol cause replication block and therefore are believed to have a strong cytotoxic effect. Ionising radiation induces mutation and chromosomal aberrations in cells, which are believed to lead to cancer and loss of neural function in humans. Humans are daily exposed to low doses of radiation during air travel or from radon in homes and areas of low-level contamination.

Energy from ionising radiation, such as X-rays and gamma rays, is transmitted in the water surrounding the DNA molecule in such a way that between 2 to 5 radical pairs are generated within a radius of 1 to 4 nm. Clustered multiple damaged sites induced by ionising radiation, have been observed, most of them being modified bases rather than DNA strand breaks. The complexity of radiation-induced clustered DNA-damage depends on the ionising density of LET (Linear Energy Transfer) radiation. Ionising radiation was hypothesised to produce clustered damage, and clusters spanning several kilobase pairs to a few base pairs were modeled.

Irradiation of DNA in non-radioquenching solution with gamma rays was shown to induce 1 double strand break to ~0.5 Nth protein-recognised oxidised pyrimidine cluster: 1.5 Nfo protein-recognised abasic cluster: 2 Fpg protein-recognised oxidised purine clusters. Use of the polyamines putrescine to measure abasic clusters refractory to Nfo protein cleavage reveals twice as many abasic clusters in -irradiated DNA as does Nfo, suggesting that even more clusters are produced than these values indicate. The dose-response relations in a log-log plot are straight lines with slopes near 1, showing that clusters are formed by a single radiation hit.

All DNA lesions are not equally frequent components of clusters, with about 15% of oxidised pyrimidines, oxidised purines and abasic sites in clusters, but only about 8% of the strand breaks are in DSBs. Since double strand break yields are ~100 times higher in the absence of radical scavengers fewer clusters are expected in DNA irradiated in the presence of a radical scavenger such as Tris.

Indeed the absolute levels of bistranded cluster levels are significantly reduced in radioscavenging solution, and, surprisingly, the ratios of specific cluster types also changed strikingly. Cluster levels in human cells exposed to 50 kVp X-rays shows that X-rays induce 1 DSB: 0.75 Nfo-abasic cluster: 1 Fpg-oxidized purine: 0.9 Nth-oxidized pyrimidine cluster. In cells, non-DSB clusters are at least ~70% of the complex damages. Clusters are postulated to be critical because they may be more difficult to repair than random lesions. In contrast to randomly located oxidised bases, clustered modified bases are within half a turn of the double helix, i.e. 5 nucleotides and some of them on the two strands, therefore the excision repair pathways would have difficulty in removing clustered lesions.

It is anticipated that the excision/incision of oxidatively damaged bases, including AP sites, on both strands will, if not tightly regulated, either inhibit certain steps of repair or produce double strand breaks and thus be lethal for the cells. Studies of purified enzymes acting on oligonucleotides with defined lesions at specific relative spacing on opposing strands indicate that clusters may comprise non-repairable, highly repair-resistant and pre-mutagenic damage, and that lesion spacing and polarity are important.

However, the precise repair mechanisms for the clustered lesions are so far very poorly understood. *E. coli* generate high levels of DSBs during repair after irradiation, and the repair-generated DSBs likely result from abortive cluster repair. Although rodent cells exposed to high doses (500 Gy) also increase DSB levels, their origin and whether they are generated at low doses was not clear.

BASE EXCISION REPAIR

Oxygen radicals generate mostly non-bulky DNA lesions, most of them are substrates for BER enzymes. Only few oxidative DNA lesions such as cyclopurine and pyrimidopurinone are substrates for nucleotide excision repair (NER) enzymes. Key enzymes of the BER pathway are DNA-glycosylases. They remove damaged and mispaired bases from DNA by cleavage of the *N*-glycosylic bond between the abnormal base and deoxyribose, leaving an abasic site in DNA. Most DNA glycosylases are highly specific and can excise various types of modified bases. There are two types of DNA glycosylases: mono- and bifunctional.

The mono-functional ones cleave the *N*-glycosylic bond releasing the modified base and generating an AP site as a final product which in turn is recognised by an AP endonuclease. The bifunctional ones cleave the *N*-glycosylic bond, liberate the modified base and in a concerted manner cleave the phosphodiester bond 3' to the resulting AP site by a or - elimination mechanism (-lyase activity) generating a single-strand break with 3'-phosphate and 3'-phosphoglycolate (PGA) extremities respectively.

Then the DNA backbone next to the abasic site or 3'-phosphate/PGA terminus is cleaved by an AP-endonuclease allowing a DNA polymerase to fill the gap before DNA ligase reseals the DNA. Experiments with *E. coli* mutants deficient in BER pathway demonstrated the crucial role of DNA glycosylases and AP-endonucleases in protecting cells from mutagenic and cytotoxic effects of free radicals. In human cells BER proceeds via two alternative pathways, either 'short-patch' DNA polymerase -dependent pathway which involves the replacement of a single nucleotide or 'long patch' PCNA-dependent pathway which involves the replacement of up to six nucleotides.

The former pathway was reconstituted *in vitro* with the purified human proteins uracil-DNA glycosylase (UDG), AP endonuclease 1 (APE1), DNA polymerase (pol), the scaffold protein XRCC1 and DNA ligase (I or III). Long-patch pathway can be achieved with AP endonuclease, Pol and PCNA. The antibodies directed against PCNA totally suppress repair patches longer than one nucleotide. Reconstitution *in vitro* of long-patch repair showed that DNase IV/FEN 1 is required in addition to proteins implicated in short-patch pathway and that PCNA greatly stimulates the reaction. Either DNA polymerase/ or DNA polymerase have been shown to perform the synthesis step in this sub-pathway. Recently, it has been demonstrated that removal of 8-oxoG could be achieved when using only hOGG1, Ape1, pol and DNA ligase I.

BASE EXCISION REPAIR IN PROKARYOTES

Mechanistic studies of both classes of DNA glycosylases led to formulation of a model of unified catalytic mechanism. Monofunctional DNA glycosylases cleave the glycosidic bond by a hydrolytic mechanism, activating a water molecule to attack the C1' of the damaged base, resulting in AP site as a final product. In contrast, the DNA glycosylases/AP lyases employ a nucleophilic group in the enzyme to perform attack at C1'.

In all DNA glycosylases/AP lyases so far characterised, an amino group has been implicated as a nucleophile. As a result of nucleophilic attack by amino group, a covalent imino intermediate (Schiff base) is formed between the C1' of the lesion and the enzyme. By abstracting the C2'-*pro-S* proton, the enzyme initiates electron rearrangement that leads to the release of the 3'-phosphate. The Schiff base is then hydrolysed, releasing the enzyme and leaving a nick with an,-unsaturated aldehyde at the 3'-end and a phosphate at the 5'-end of the DNA. Upon completion of -elimination, some enzymes such as the Fpg protein abstract proton at the C4' and promote a -elimination.

The resulting structure is a single-base gap with phosphates at both 3'- and 5'-ends. Sodium borohydride and cyanoborohydride have been used to show that the reaction involves a Schiff base intermediate. When the reduction reaction occurs in the glycosylase-DNA complex, the enzyme becomes irreversibly crosslinked to DNA. *The Fpg protein (formamidopyrimidine-DNA glycosylase):* Among the DNA glycosylases, the *E. coli* Fpg protein (formamidopyrimidine-DNA glycosylase/MutM protein) has been one of the most extensively studied. The gene coding for the Fpg protein was cloned and the physical and enzymatic properties of the protein established. It is a globular monomer of 30.2 kDa of 269 amino acids. *In vitro,* the Fpg protein excises a broad spectrum of modified purines, particularly 2,6-diamino - 4 - hydroxy - 5N - methylformamidopyrimidine (Fapy) and 7,8-dihydro-8-oxoguanine (8-oxoG) residues.

Moreover, the Fpg protein is also able to excise various pyrimidine oxidation products such as 5-hydroxycytosine and 5-hydroxyuracil, the ring fragmentation product of thymine (RT), thymine glycol and 5,6-dihydrothymine. The Fpg protein is a bifunctional DNA glycosylase, endowed of an AP lyase activity that incises DNA at abasic sites by a --elimination mechanism and an activity excising 5'-terminal deoxyribose phosphate (dRPase). *In vivo,* the Fpg protein has an antimutator effect preventing G/CT/A spontaneous transversion. It acts in concert with the MutY protein. The *fpg mutY* double mutant CC104 has an extreme mutator phenotype that can be reversed by plasmids carrying the fpg gene.

Interestingly, expression of the bacterial fpg gene in mammalian cells reduces the mutagenicity of -rays but has no effect on survival. The Fpg protein in its COOH-terminus contains the zinc-finger motif (Cys-X_2-Cys-X_{16}-Cys-X_2-Cys-X_2-COOH) that is mandatory for Fpg binding to DNA and to its enzymatic activities.

The active site of Fpg protein is located within the first 73 amino acid residues of the amino terminus. The targeted mutagenesis of conserved amino acid residues was used to elucidate the mechanism of enzymatic catalysis. It was found that conserved residues lysine 57 (K57G), lysine 155, glutamates 2, 5, 131 and 173 and proline 2 (P2G) dramatically reduce the cleavage of oxidised bases such as 8-oxoG. Recently, major progress has been achieved in solving the three-dimensional structure of the Fpg protein from various origins.

Two Fpg proteins have been isolated from the highly radioresistant bacteria *Deinococcus radiodurans*. Both excise Fapy and 8-oxoG residues and present a -lyase activity. In addition one excises thymine glycols. The genes have been cloned and the detailed specificities established using pure proteins. *The Nth protein (Endonuclease III):* The Nth protein was initially identified as an endonuclease that specifically cleaves DNA damaged by X-rays, UV light and free radicals. It is a DNA glycosylase with a broad substrate specificity, excising ring-saturated, ring-opened, and ring-fragmented pyrimidines, such as thymine glycol, 5,6-dihydrothymine, 5-hydroxy-6-hydrothymine,5,6-dihydrouracil, alloxan, 5-hydroxy-6-hydrouracil, uracil glycol, 5-hydroxy-2'-deoxy-cytidine, 5-hydroxy-2'deoxyridine, -ureidoisobutiric acid and also -R-hydroxy--ureidoisobutiric acid, a fragmentation product of the 5R-thymidine C5-hydrate.

Unexpectedly, it has been demonstrated that Nth is implicated in the repair of 8-oxoG, since it removes 8-oxoG from 8-oxoG/G mispairs and the triple mutant *fpg nth nei* displays increased spontaneous G/CC/G transversions as compared to single and double mutants in these three genes.

This leads to the proposal of a new model in which Nth and Nei repairs 8-oxoG in nascent and transcriptionally active DNA. The nicking activity at abasic sites of Nth is due to its AP lyase function. The phosphodiester bond cleavage occurs via -elimination (the protein does not cleave at reduced AP sites) generating 5' ends bearing 5'-phosphate and 3' ends with 2,3-unsaturated abasic residue 4-hydroxy-2-pentanal. This terminus requires further processing to be used for DNA repair synthesis. The gene encoding for the Nth protein, the *nth* gene, was cloned and protein purified to homogeneity. The three dimensional structure of Nth was also established. The Nth protein is a monomeric protein of 23.4 kDa (211 aa).

It is an elongated protein with a cleft separating two similar size domains: a continuous domain formed by six-- helices and a second domain formed by three C-terminal - helices and the N-terminal helix. The C terminal loop contains an iron-sulphur centre (4Fe-4S) that is anchored to the protein by a Cys-X6-Cys-X2-Cys-X5-Cys sequence. This cluster has a structural function in positioning basic residues for DNA-binding and it is also present in the MutY protein.

E. coli nth mutants deficient in the Nth protein does not show apparent phenotype. They are not sensitive to X-rays,

H_2O_2 or other agents that produce ring saturation and fragmentation products of pyrimidines in DNA. However the double mutant *nth nei* exhibits a strong spontaneous mutator phenotype and it is hypersensitive to ionising radiation and H_2O_2.

The Nei protein

The capacity of *E. coli nth* mutants defective in most of the activity to excise thymine glycol, allowed the identification of a new activity excising this oxidised base, that was named endonuclease VIII or Nei protein. It has been purified from *E. coli* extract lacking the Nth protein. Similarly to the Nth protein, Nei removes the oxidised forms of thymine such as thymine glycol, dihydrothymine, -ureidoisobutyric acid, and urea residues when present in DNA and cleaves phosphodiester bond at AP sites.

The cloned *nei* gene encodes a 263 amino acid polypeptide with a striking similarity to the Fpg protein. The catalytic site's residues are highly conserved in these two proteins, however, their substrate specificities are different. Characterisation of apparently homogenous Nei protein has shown that it can equally remove the oxidised forms of cytosine such as 5-hydroxycytosine and 5-hydroxyuracil. More recently, genetic evidence for Nei implication in the prevention of spontaneous GT transversions was confirmed since Nei was found to cleave 8-oxoG efficiently when opposite to A and G.

Furthermore, extended the substrate specificity of the Nei protein and have demonstrated that Nei and Nth significantly differ from each other in terms of kinetic although they share common substrate specificity. The crystal structure at 1,25 Å of the Nei covalent intermediate complex with DNA largely confirms that its structure is similar to that of Fpg.

MONOFUNCTIONAL DNA GLYCOSYLASES

AlkA protein

The main substrates for the *E. coli* AlkA and Tag I proteins are alkylated bases. The Tag I protein is a major DNA glycosylase removing 3-methylpurines residues. However, the AlkA protein in contrast to Tag I is also involved in repair of oxidative DNA damage. Hypoxanthine (HX) is generated in DNA by spontaneous deamination of adenine and also by the free radical nitric oxide. HX residues in DNA are mutagenic since they can pair with cytosine, generating AT to GC transitions after DNA replication.

Hypoxantine-DNA glycosylase releases HX from DNA containing dIMP. Cloning the gene coding for the HX-DNA glycosylase in *E. coli*, it was shown that the 3-methyladenine-DNA glycosylase coded by the alkA gene carries the HX-DNA glycosylase activity. ANPG, APDG and MAG proteins, the human, rat, and yeast functional homologues respectively of AlkA protein excise HX residues when present in DNA, the mammalian enzymes being most efficient. The AlkA protein also catalyses the excision of ethenobases N^2,3-ethenoguanine and 1,N^6-ethenoadenine.

The 5-formyluracil (5-foU) residue, an oxidised thymine lesion is a major DNA damage induced by ionising radiation. It induces A/T to G/C transition and is repaired in *E. coli* by the AlkA protein. In addition, AlkA protein excise RT residues when present in DNA with an apparent K_m=170 nM. Thus the AlkA protein has very broad substrate specificity. The three dimensional structure of AlkA reveals a compact globular protein with a prominent hydrophobic cleft on its surface and three equal-sized domains.

Uracil DNA glycosylase

Base excision repair was first established for uracil. Four sub-families of uracil-DNA glycosylase (UDG) have been identified. The best studied family of UDGs is *E. coli* Ung protein. Ung is specific for uracil and it is present in a wide range of living organisms from bacteria, lower to higher eukaryotes, DNA viruses and human. So far UDG has not been implicated in the repair of oxidative DNA damage. The second family corresponds to the mismatch-specific uracil-DNA glycosylase (MUG) identified in some eukaryotes and in several prokaryotes. MUG excises thymine from G/T mismatches in DNA and it is also active on mispaired uracil in G/U pair. The crystal structure of Ung and Mug proteins show that they are structurally similar, despite low sequence homology.

Another sub-family has been characterised from thermophilic archea and several bacteria. These double strand UDGs (dsUDG/DUG) are able to remove uracil from double strand DNA either in U/A and U/G context. The fourth sub-family, initially identified in vertebrates is represented by the human single-strand mismatch-specific uracil-DNA glycosylase (hSMUG1). However hSMUG1 is more active on double-stranded than on single-stranded DNA, it prefers uracil opposite A and G.

The 5-hydroxymethyluracil residue is a product of oxidation and subsequent deamination of 5-methylcytosine. Teebor's group identified the mammalian 5-hydroxy-methyluracil DNA N-glycosylase activity as SMUG1. A fifth sub-family has been characterised recently and is represented by *pa*-UDGb isolated from *Pyrobaculum aerophilum*. *pa*-UDGb has broad substrate specificity. It can remove uracil and also hypoxanthine when present in DNA.

Among the members of the UDG family several enzymes are implicated in the repair of exocyclic DNA adducts which are highly mutagenic oxidative DNA lesions. The *E. coli* MUG protein is able to remove 3,N^4-ethenocytosine, 8-(hydroxymethyl)-3,N^4-ethenocytosine and 1,N^2-ethenoguanine when present in DNA.

MutY

The *E. coli mutY* gene encodes a DNA glycosylase of 39 kDa, which excise A opposite G and C. MutY also removes A opposite to 8-oxoG and 7,8-dihydro-8-oxoadenine (8-oxoA). The protein shows significant sequence homology to Nth and is also an iron-sulphur protein which contains the conserved (4Fe-4S) cluster/HhH domain. The protein was extensively characterised using various approaches including site-directed mutagenesis. However, there is still controversy regarding the AP lyase activity of the MutY protein. Proteolytic cleavage of

MutY generates two fragments of 26 kDa and 13 kDa. The 26 kDa fragment corresponds to the catalytic core domain, it retains normal DNA binding and adenine-DNA glycosylase activity but lacks its activity against A/8-oxoG.

The specificity for 8-oxoG residues is due to the C-terminal domain of MutY. It is thought that MutY plays an important role in the maintenance of genome integrity following oxidative DNA damage. The presence of AP endonucleases (Xth and Nfo) greatly enhances the excision rate of A when opposite to G by MutY. These data suggest that the MutY-DNA complex interacts with Nfo and Xth proteins *in vivo*. The crystal structure of MutY catalytic core domain (cMutY) has been solved. cMutY is an all- protein that retains the canonical bilobal architecture of Endonuclease III.

The compression of intra-strand phosphate distance by HhH and pseudo HhH domains permit the DNA bending and the nucleotide flipping. The flipped nucleotide is recognised by specified residues of the active side pocket located between the two domains. Crystal structures and mutagenesis results define that MutY cleaves the N-glycosylic bond through a hydrolytic mechanism requiring Asp138, with uncoupled, inefficient AP lyase activity due to Lys142.

AP ENDONUCLEASES: XTH AND NFO

Abasic sites (AP sites) occur in DNA through spontaneous depurination/depyrimidination, by the action of ROS and as secondary lesions after excision of modified bases by DNA glycosylases. AP sites have miscoding properties since the replication machinery incorporates preferentially adenine opposite to AP site. This observation was later named 'A rule'. The occurrence of an enzyme recognising AP sites in *E. coli* was first described in the seminal work of Verly.

Two types of enzymes recognise and excise DNA at AP sites: AP endonucleases such as Nfo and Xth proteins act on this lesion by a hydrolytic mechanism whereas enzymes carrying a -lyase activity such as Fpg, Nei and Nth proteins incise the AP site via a - and elimination mechanism respectively.

THE NFO PROTEIN (ENDONUCLEASE IV)

The Nfo protein is an EDTA-resistant AP endonuclease that represents only 5% of the total AP endonucleasse activity in *E. coli* wild type. It is inducible by oxidative stress and is under the control of the soxRS system. The gene coding in *E. coli* for Endonuclease IV, *nfo,* has been cloned, and the protein characterised. It is a monomer of 30 kDa having several activities: AP endonuclease, 3'-phosphatase and 3-phosphoglycoaldehyde diesterase activities.

It has been suggested that these 3' repair activities clean the ends, a prerequisite for DNA synthesis. *E. coli nfo* mutants deficient in the product of the *nfo* gene are extremely sensitive to the lethal effects of oxidative agents such as bleomycin and t-butyl-hydroperoxide. In addition, this mutation enhances the sensitivity of *xth* mutants to H_2O_2, and alkylating agents.

The double mutants *xth nfo* are also sensitive to -radiation. We will discuss below the role of the Nfo protein and its yeast homologue Apn1 in the nucleotide incision repair pathway (NIR). The high-resolution structure of Nfo protein has been solved. It shows that Nfo is an protein arranged as a $_{88}$-barrel. This structure, well suited for the binding to large molecules such as DNA, contains three Zn^{2+} ions at the active site which are critical for the activity.

The triple Zn centre is ligated to protein by conserved residues that cluster at the centre of the crescent-shaped deep pocket. The Nfo protein detects AP site by insertion of its side chains into the DNA minor groove, it flips the target AP site and the opposite nucleotide out of the DNA base stack to produce a 90° bend in the DNA.

The Xth Protein

The Xth protein, coded for by the xth gene, originally identified as a 3'5' exonuclease active on double-stranded DNA and a 3'-phosphatase, is the major AP endonuclease of *E. coli*: over 80% of the total AP endonuclease activity in wild type. This protein is also active on single stranded DNA. Beside its exonuclease and 3'-phosphatase activities it has 3' repair diesterase and ribonuclease H activities. The 3' termini generated by Xth are normal nucleotides with 3' hydroxyl group that are effective primers for DNA polymerases.

The *xth* mutants are extremely sensitive to H_2O_2, and to oxidative DNA damage generated by near-UV light. The crystal structure of the Xth protein has been solved, and it has been assumed that the base opposite to the abasic site plays an important role for the recognition of the lesion.

Mechanism of action and three-dimensional structure of bifunctional DNA glycosylases: the Fpg case: It was hypothesised that DNA glycosylase/AP lyases use a mechanism involving the nucleophilic attack on the C1' prime of the modified deoxynucleoside targeted for excision, thus displacing the aberrant base and forming a transient intermediate (Schiff base) with the C1'-deoxyribose moiety. The amino acid sequence of the Fpg protein begins with a Met-1 which is processed, thus Pro-2 being the N-terminal.

The N-terminal proline (P2) residue of the Fpg protein, a highly conserved residue, was shown to be linked with DNA containing 8-oxoG residues, suggesting that proline is the nucleophile initiating the excision of 8-oxoG. Site-directed mutagenesis demonstrated the mandatory role of the N-terminal proline residue in the 8-oxoG-DNA glycosylase activity of the Fpg protein *in vitro* and *in vivo,* as well as in its AP lyase activity upon pre-formed AP sites but less of a role in Fapy-DNA glycosylase activity.

The role of the conserved lysine 57 and 155 (K57 and K155) residues upon the various catalytic activities of the Fpg protein was examined by targeted mutagenesis. The lysine 57glycine (FpgK57G) mutant protein had a dramatically reduced (55-fold) DNA glycosylase activity for the excision of 8-oxoG residues. The FpgK57G protein was poorly effective in the formation of Schiff base complex with 8-oxoG/C DNA. However, the mutant could partially restore the ability to prevent spontaneously induced G/CT/A transversions in *E. coli* BH990 (*fpg, mutY*) cells.

The DNA glycosylase activity of FpgK57G using FapyGua residues as the substrate was comparable to that of the wild type enzyme. These results suggested that K57 could participate either in a direct interaction with the C_8 oxygen of the 8-oxo purines or in the opening of the furanose ring, in the first step of the model proposed for the mechanism of action of the Nth protein.

Effect of mutations of conserved glutamic and aspartic acid residues to glutamines and asparagines, have been studied. While the Asp to Asn mutants had no effect on the incision activity on 8-oxoG DNA, several of the substitutions at glutamates reduced Fpg activity on the 8-oxoguanosine DNA, with the E3Q and E174Q mutants being essentially devoid of activity. Unexpectedly, the AP lyase activity of all of the glutamic acid mutants was slightly reduced as compared to the wild-type enzyme. Furthermore, it has been shown that lysine 57 but not proline 2 is crucial for catalysis of oxidatively damaged pyrimidines.

Sugahara and co-workers determined the crystal structure of Fpg homologous protein from an extreme thermophile, *Thermus thermophilus* HB8 (HB8-Fpg) at 1.9 Å resolution. It reveals that the Fpg protein molecule is composed of two distinct domains connected by a flexible hinge.

More recently, Castaing and colleagues resolved the structure of a non-covalent complex between the *Lactococcus lactis* Fpg and a 1,3-propanediol (Pr) abasic site analogue-containing DNA. They have shown that Fpg pushes out the Pr site from the DNA double helix, recognising the cytosine opposite the lesion and inducing a 60 degree bend of the DNA.

Finally, the structure of a trapped catalytic intermediate of *E. coli* Fpg protein has been determined at 2.1 Å resolution. In agreement with the structure of HB8-Fpg the *E. coli* Fpg is a bilobal protein with a wide negatively charged DNA-binding groove.

Highly conserved residues Lys-57, His-71, Asn-169 and Arg-259 are involved in binding the phosphodiester backbone of DNA, which is sharply kinked at the lesion site.

Conserved residues Met-74, Arg-110 and Phe-111 are inserted into DNA helix to fill the void in DNA after nucleotide eversion. A deep hydrophobic pocket in the active site is positioned to accommodate the damaged base.

BASE EXCISION REPAIR IN EUKARYOTES

Yeast

Monofunctional DNA glycosylase: spMYH: The *E. coli* MutY homolog was identified in *Schizosaccharomyces pombe* by a homology search in genome databases. The spMYH gene encodes for a 461 amino acids protein which is highly homologous to MutY and human MYH. spMYH possesses the HhH motif and the [4Fe-4S] DNA binding cluster domain found in Nth/MutY.

It has a DNA glycosylase and probably an AP lyase activity. SpMYH incises A opposite G (A/G), A/8-oxoG, 2-aminopurine/G and A/2-aminopurine containing DNA. Cells deleted for the spMYH gene (*spMyH*) have a mutator phenotype and are more sensitive to oxidative agents, such as hydrogen peroxide, than wild type cells.

These data show that spMYH is implicated in the avoidance of 8-oxoG induced mutations and it has an important role in the protection against oxidative stress. Interestingly, similar to human MYH (hMYH), spMYH also interacts directly with PCNA. This interaction is fundamental for the biological function of spMYH in the control of mutation avoidance since the mutation rate of *spMyH* expressing hMYH is reduced but it stays unchanged when *spMyH* cells express a mutant hMYH protein unable to interact with PCNA.

Bifunctional DNA glycosylases S. cerevisiae *yOGG1, Ntg1 and Ntg2: Saccharomyces cerevisiae* 8-oxoguanine-DNA glycosylase (yOGG1) is the yeast counterpart of the *E. coli* Fpg protein. It was identified by functional suppression of the *E. coli fpg mutY* mutator phenotype, leading to the concept that eukaryotes use the DNA glycosylase pathway to repair oxidised purines in DNA. Comparison of amino acid sequences of the yOGG1 protein and *E. coli* Fpg protein does not reveal any obvious homology. The yOGG1 protein has associated DNA glycosylase/AP lyase activity and cleaves abasic sites via -elimination. In contrast to the bacterial enzyme, the yOGG1 protein repairs 8-oxoG only when paired with pyrimidines.

The yOGG1 also removes Fapy guanine-derived residues from DNA but less efficiently than 8-oxoG and -elimination was not observed. The major difference between Fpg and yOGG1 in terms of substrate specificity is that Fpg also removes adenine-derived Fapy lesions and oxidatively damaged pyrimidines. The yOGG1 is not an essential gene, as its disruption had no effect on the viability of a *ogg1* haploid mutant but *ogg1* cells display a mutator phenotype characterised by an augmentation of G/CT/A transversions. Like *fpg* cells, *ogg1* haploid mutants do not show particular sensitivity to H_2O_2. These data suggest that yOGG1 is the functional homologue of *E. coli* Fpg.

Based on the fact that the helix-hairpin-helix (HhH) DNA-binding motif was found in many DNA binding proteins, searches in databases for sequences containing this motif led to identification of the NTG1 gene on chromosome I of *S. cerevisiae*. The NTG1 gene encodes a 45 kDa protein, Ntg1p, which is homologous to the *E. coli* endonuclease III (Nth) but lacks the (4Fe-4S) cluster DNA binding domain found in the other members of this family.

Ntg1p is a DNA glycosylase/AP lyase which removes thymine glycols and unexpectedly releases formamidopyrimidine residues from DNA with a high efficiency comparable to that of the *E. coli* Fpg protein. Ntg1p also excises other lesions generated by oxidative stress such as 5,6-dihydrouracil, and 5-hydroxy-purines and 8-oxoG only when paired with guanine. Targeted disruption of the NTG1 gene results in viable cells. The Ntg1 protein localises both in the nucleus and in the mitochondria and is induced by cell exposure to DNA-damaging agents purified and characterised a second *S. cerevisiae* Ogg-like protein, called Ogg2, which preferentially acts on 8-oxoG/G base pairs. Later it turned out that Ogg2 is identical to Ntg1.

NTG2 was identified during analysis of *S. cerevisiae* genome as a second gene homolog to *E. coli* Nth in yeast. The NTG2 gene is localised on chromosome XV and codes for a 43.7 kDa protein, Ntg2p, which contains (4Fe-4S) cluster DNA binding domain also present in the *E. coli* Nth protein. The substrate specificity of Ntg2p is similar but not identical to Ntg1p; it cannot incise 8-oxoG mispaired with any of the four DNA bases. Recent results show that Ntg2p, but not Ntg1p, is implicated in the repair of certain degradation products of 8-oxoG such as 8-hydroxydeoxy-guanosine (8-OH-dG).

The targeted disruption of NTG2 results in viable cells but, similarly to *ntg1⁻* cells, greatly increases the rate of spontaneous and H_2O_2-induced mutations. Ntg2p is a nuclear enzyme constitutively expressed in cells. Interestingly, Ntg2p interacts with the DNA mismatch repair protein Mlh1p. Ntg1p and Ntg2p are both required for repair of spontaneous and induced oxidative DNA damage.

S. cerevisiae *AP endonucleases Apn1 and Apn2:* Homologs of *E. coli* Xth and Nfo have been identified in eukaryotes. The yeast APN1 gene encodes AP endonuclease I (Apn1) homologous to *E. coli* Nfo protein. This 41.4 kDa protein is the major AP endonuclease in yeast, accounting for >90% of total AP endonuclease activity. Apn1 has AP endonuclease, 3'-diesterase and 3'-phosphatase activities.

Like the *E. coli* homolog, Apn1 is a metalloenzyme excising 3'-phosphoglycoaldehyde, 3'-phosphoryl groups, and 3'-, unsaturated aldehydes. Yeast mutant lacking Apn1 (*apn1*) is viable but it is hypersensitive to both oxidative (H_2O_2 and t-butylhydroperoxide) and alkylating (methyl- and ethylmethane sulfonate) agents, it has 6- to 12-fold higher rate of spontaneous mutation than wild-type. This mutator phenotype is mainly characterized by a 60-fold increase in A/T to G/C transversion rate. The second yeast AP endonuclease (Apn2/ETH1) was identified by blast homology search with the sequences of *E. coli* Xth and human Ape1 in the *S. cerevisiae* genome.

Apn2 is a 520 amino acid protein, its expression (at the mRNA level) is induced by DNA damage. Apn2 has AP endonuclease, 3'-phosphodiesterase and 3'5' exonuclease activities, it can remove 3'-phosphate and 3' phosphoglycolate termini produced by H_2O_2. Interestingly, the 3' phosphodiesterase and the 3'5' exonuclease activities of Apn2 are 30-40-fold more active than its AP endonuclease activity. Apn2 could be an important factor in the repair of oxidative DNA damage.

Yeast lacking apn2 (*apn2*) alone are viable and do not show a particular phenotype.

However *apn1 apn2* cells are remarkably sensitive to DNA damaging agents such as H_2O_2, MMS and phleomycin D1 and they present an increase of spontaneous mutation rate. Apn2 contains a carboxy-terminal domain which is absent in the other members of the family (Xth/Ape1). Deletion of this domain does not affect the enzymatic activity of Apn2 *in vitro* but the truncated protein cannot remove AP sites *in vivo*. These data strongly suggest that this domain is indispensible for protein-protein interaction and that Apn2 protein functions *in vivo* as a part of multiprotein complex.

Interestingly, it has been shown that the human homolog of Apn2 protein, Ape2, interacts with PCNA.

MAMMALS (HUMAN, RODENT)

Bifunctional mammalian DNA glycosylases OGG1, NTH1 and NEH1 hOGG1: Several groups reported the molecular cloning of the cDNA of the human and murine 8-oxoguanine DNA glycosylase (OGG1). The amino acid sequence of hOGG1 showed 33% identity and 54% similarity to *S. cerevisiae* OGG1 and conserved HhH/PVD motif which is present in endoIII/ MutY/AlkA superfamily of DNA glycosylases. The human OGG1 gene, located on chromosome 3, codes for seven alternatively spliced forms of mRNA.

These transcripts were classified as type 1 and 2 depending on their last exon. The main type 1 and 2 transcripts encode, respectively, a 36 kDa nuclear and a 40 kDa mitochondrial polypeptide. These two proteins differ by their C-terminus.

Both forms possess the same catalytic activity. Purified hOGG1 protein, similarly to *E. coli* Fpg protein, acts as a DNA glycosylase towards duplex DNA containing 8-oxoG/C base pair, Fapy and has associated AP lyase activity.

Crystal structure data at 2.1 Å resolution of the core domain of hOGG1 complexed with a 8-oxoG/C containing oligonucleotide was reported by Verdine's laboratory. This structure reveals that hOGG1 recognizes in the DNA helix both

8-oxoG (using amino acids F319, Q315, G42, C253) and the cytosine opposite 8-oxoG (using amino acids N149, R154, R204, Y203). The modified base is fully extruded from the helix and inserted into an extra-helical active-site pocket on the enzyme.

Interestingly, mutations of residue, which recognises the cytosine, does not alter the activity on 8-oxoG/C but significantly increases the activity toward 8-oxoG opposite other bases than cytosine. Physiologically, such mutants become both pro-mutagenic by repair of 8-oxoG/A and anti-mutagenic by repair of 8-oxoG/C. Study of search intermediates of hOGG1 protein in the presence of high molecular weight DNA by atomic force microscope shows that enzyme scans DNA at undamaged sites by inducing drastic kinks.

The base excision repair pathway for 8-oxoG involves the resynthesis of a single nucleotide at the lesion site as shown by *in vitro* repair assays with mammalian cell extracts and more recently by reconstitution *in vitro* of this pathway by using human purified proteins. These data show that the BER short-patch pathway is predominant for the repair of major oxidative DNA damage, 8-oxoG. Activity of hOGG1 is greatly stimulated by Ape1 which binds to AP sites generated by the DNA glycosylase and enhances the enzymatic turnover. This observation suggests that Ape1 might preclude the AP lyase activity of OGG1. The finding that the dRP lyase activity of DNA polymerase is required in repair of oxidative lesions by mammalian cell extracts supports this mechanism.

The nuclear hOGG1 is associated with chromatin and the nuclear matrix during interphase and with condensed-chromatin during mitosis. The fraction of hOGG1 bound to chromatin is phosphorylated on a serine residue, and the kinase responsible for this post-translational modification of hOGG1 might be protein kinase C. Homozygous *ogg1-/-* null mice were generated by targeted disruption. These mice are viable and, despite an increase of potentially mutagenic DNA lesions in their genome and mitochondria, they do not display any particular phenotype.

The *ogg1-/-* null mice have an elevated spontaneous mutation rate in non- or slowly proliferating tissues with high oxygen metabolism, such as liver, but they do not develop malignancies. These data are at variance with several reports which associate mutations and polymorphism of the OGG1 gene, in particular the Ser326Cys polymorphism, with increased risk of cancer. Although the Ser326Cys polymorphism is slightly, or not associated with altered OGG1 activity, it might affect phosphorylation at this serine residue leading to adverse effects.

hNTH1: A human homolog of the *E. coli* Nth protein (endonuclease III) has been purified and cloned. This gene, called hNTH1 (human Nth homolog 1), encodes a 34.3 kDa protein that shares extensive homology with Nth including the conserved HhH-motif and the (4Fe-4S) cluster loop motif. hNTH1 protein acts as DNA glycosylase with an associated AP lyase activity, and has substrate specificity similar to the *E. coli* homolog towards damaged pyrimidine derivatives that result from ring saturation, ring fragmentation, ring contraction and unexpectedly ring-opened purine residues (Fapy). hNTH1 acts preferentially on 5-hydroxycytosines and AP sites when they are situated opposite to guanine and removes 8-oxoG opposite G. hNTH1 seems to be an exclusively nuclear protein and its expression is regulated during the cell cycle, showing a maximum in S-phase.

Homozygous *mNth1-/-* mutant mice do not have any detectable phenotype defect. Sensitivity of *mNth1-/-* mutant embryonic cells to H_2O_2 and menadione was not changed as compared to wild-type cells. Moreover, thymine glycol induced by ionising radiation is repaired, though more slowly than in wild type. The absence of a significant phenotype could be explained by the discovery of two novel thymine-glycol DNA glycosylases in *mNth1* mutant cells called TGG1 and TGG2. These ~40 kDa proteins localise in mitochondria and nucleus, respectively.

Thus, these data suggest the existence of several back-up DNA glycosylases in mammals to remove thymine glycol. The redundancy of such enzymatic activity highlights the importance of BER pathway for counteracting oxidative pyrimidine damage.

Although there is no definite evidence that BER enzymes act as multiprotein complexes, several reports suggest that addition of proteins can stimulate damage recognition and lesion processing. The NER enzyme XPG stimulates the hNTH1 activity *in vitro*. These data support the hypothesis that neurodegeneration in XP-patients could be due to defects in the repair of oxidative DNA damage. A yeast two-hybrid screen for other interacting partners of Nth1 led to the isolation of the DNA-binding protein B (DbpB)/Y box binding protein (YB-1). YB-1 greatly stimulates the hNTH1 activity. Altogether these data suggest that the BER pathways could be regulated by a number of non-BER proteins.

hNEH1: Human homologs of *E. coli* Nei/MutM were identified by database search of the genome and named hNEH1 and 2 (human Nei Homolog 1 and 2). hNEH1 was biochemically characterized, it encodes a 44 kDa protein which excises Fapy, oxidised purines and 8-oxoG when present in DNA. hNEH1 shows a tissue-specific expression with an S-phase specific increase at both RNA and protein level. Tissue-specific mRNA levels of OGG1 and NEH1 are distinct. These data suggest that the activities of OGG1 and NEH1 are not redundant at the tissue level and NEH1 might be involved in the replication-coupled repair of oxidative DNA damage.

Monofunctional DNA glycosylases ANPG, hTDG, MYH

ANPG: The ANPG protein is the human counterpart of the *E. coli* AlkA protein. This protein releases 3-meA and 7-meG from methylated DNA, but it is also able to excise hypoxantine and 1,N^6-ethenoadenine more efficiently than the AlkA protein. In contrast to the AlkA protein, the ANPG protein does not release 5-formyluracil and RT residues from DNA, however, there are activities in human cell-free extracts that can excise these lesions.

It should be noted that the human protein does not share significant amino acid sequence homology with the bacterial AlkA. Recently, it has been shown that ANPG efficiently excises 1,N^2-ethenoguanine (1,N^2-G) when present in DNA. Immunofluorescent staining for the ANPG protein in normal breast cells showed its nuclear localisation. Two alternatively spliced transcripts of human ANPG have been isolated from human cells. The gene encoding for ANPG maps to chromosome 16. The full-length cDNA first isolated by Samson and colleagues (ANPG70, 293 AA, 1991) differs from the splice variant only in the sequence of the first 8 and 13 N-terminal amino acids respectively. Two truncated versions of human ANPG have also been described: ANPG40 lacks the first 63 amino acids and ANPG80 the first 73 amino acids from the N-terminus, when compared to APNG70. It has been concluded that the non-conserved, N-terminal part apparently contributes little to their damage recognition and N-glycosylase activity.

Surprisingly, it has been shown that the ANPG80, which lacks 73 amino acid residues at the N-terminus, is unable to excise 1,N^2-G from the duplex oligonucleotide and that the ANPG40 has a reduced activity as compared to the ANPG70. These observations indicate that the non-conserved, N-terminal part of ANPG is essential for 1,N^2-G-glycosylase activity but dispensable for the release of A, hypoxanthine and N-methylpurines. The ANPG activity toward hypoxanthine-containing substrates is slightly activated in the presence of the hHR23 protein.

Although the core part of the ANPG70 and APDG proteins display 85% sequence identity, it seems that human enzyme repairs 1,N^2-G somewhat more efficiently than its rat counterpart. In fact, a similar difference between human and mouse ANPG in the recognition of 3-methylguanine and 7-methylguanine residues has been reported. Interestingly, the human full-length and splice variants of ANPG have a direct repeat in N-terminal amino acid sequence, which is absent in the rat and murine homologues. These observations further support the role of the non-conserved N-terminal part of mammalian ANPG in substrate specificity.

To assess the role of this repair enzyme *in vivo*, mice deficient in 3-meAde-DNA glycosylase (*APNG/Aag* null mice) have been generated. Unexpectedly, these APNG knockout mice exhibit neither any particular sensitivity to alkylating agents nor any significant increase in the spontaneous mutation rate except for splenic T lymphocytes. Furthermore, Roth and Samson have shown that *Aag-/-* myeloid progenitor bone marrow cells are more resistant than wild-type cells to alkylating agents. This result suggests that the initiation of base excision repair could be more lethal to the cell than leaving the damaged bases unrepaired.

The crystal structures of ANPG80 (a truncated form of the ANPG protein) complexed to a DNA duplex containing pyridine and A have been established. ANPG80 is a single domain protein of mixed/ structures, which forms a distinct structural group that does not resemble any other BER protein. The enzyme bends the DNA by about 20° and intercalates into the minor groove of DNA causing the abasic pyrrolidine and A to slip into the enzyme active site. The structure of the ANPG80 nucleotide-binding pocket changes little upon flipping in the A base. The structure of the base-binding pocket of ANPG reveals the lack of specific interactions for methylated bases. This fact probably contributes to its remarkable multifunctionality, providing glycosylase activity within the same enzyme for deamination, methylation and exocyclic base damage.

hTDG: The human thymine DNA glycosylase (hTDG) was first biochemically characterised for its ability to remove T mispaired with G. A more complete characterisation shows that hTDG exhibits a broad substrate specificity: it can excise T from DNA not only when mispaired with G but also from other T-containing mispairs except T/A. Moreover, the hTDG protein more efficiently removes U from a G/U but not from a U/A mispair and 3,N^4-ethenocytosine opposite to all four natural bases. The hTDG gene is localised on the chromosome 12, it codes for several mRNAs which were cloned. These cDNA encode a single 46 kDa polypeptide (410 amino acids) which is homologous to the *E. coli* MUG protein. The hTDG protein has a high affinity for AP site and stays tightly bound to it after removal of T from a G/T mispair.

Ape1 activates hTDG by increasing the dissociation rate of hTDG from AP site probably through direct interaction. The stimulation of hTDG turnover by Ape1 is responsible for the efficient processing of 3,N^4-ethenocytosine by hTDG. A second factor facilitating hTDG enzymatic turnover was recently identified. Hardeland have shown that SUMOylation of hTDG by SUMO-1 and SUMO-2/3 drastically reduces its affinity for AP sites, increases its enzymatic turn-over towards G/U and reduces its processing activity on G/T substrate. SUMOylation also enhances the stimulatory effect of Ape1 on hTDG.

Using the yeast one-hybrid screen, hTDG was found to interact with retinoic acid receptor (RAR) and retinoic X receptor (RXR) which are ligand-dependent transcription factors. Interaction between RXR, RAR/RXR and hTDG enhances the binding of the former on their responsive elements. These data suggest that hTDG has a dual function: repair enzyme and transcriptional activator. hTDG is also a transcriptional repressor since it interacts with the thyroid transcription factor 1 (TTF-1) and then strongly inhibits the expression of TTF-1 responsive genes.

Recently, hTDG has been found to interact with the transcriptional co-activator CBP/p300. The hTDG/CBP/p300 complex is competent for both BER and histone acetylation. hTDG enhances the CBP/p300 transcriptional activity and these proteins acetylate it. hTDG acetylation regulates the recruitment of Ape1 by hTDG. These data highlight the complexity of repair processes that are coupled to transcription, tightly regulated by post-translational modifications (SUMOylation and acetylation) and by protein-protein interactions.

SMUG1: The single-strand mismatch-specific uracil-DNA glycosylase (SMUG1) has been identified only in vertebrates by expression cloning. SMUG1 is a nuclear protein of 31 kDa which in fact efficiently removes uracil from double stranded DNA when paired with A and G, it also removes U from single-stranded substrates but less efficiently. Accumulation of 5-hydroxymethyluracil is associated with a cancer risk. The mammalian 5-hydroxymethyluracil DNA N-glycosylase activity is associated with SMUG1.

Like several DNA glycosylases, SMUG1 has a low turnover rate that could be due to its inhibition by AP sites.

This could explain the Ape1-dependent stimulation of the hSMUG1 activity on both single and double-stranded DNA substrates. Targeted disruption of UNG gene in mice (*ung-/-*) does not result in relevant increase of spontaneous mutation rates, contrary to bacteria and yeast *ung*⁻ mutants. It seems that hSMUG1 has an important role in the maintenance of genome stability since it represents the major uracil-DNA glycosylase activity in *ung-/-* cells.

hMYH: The human adenine-DNA glycosylase activity, called hMYH protein (human MutY homolog) was first purified from cellular extracts. Later using homology search, the genomic DNA and cDNA encoding hMYH were identified. Functional expression of hMYH in *E. coli* complements the mutator phenotype of *mutY* cells. The hMYH gene is localised on chromosome 1. It encodes a 59 kDa monofunctional DNA glycosylase which can excise A when opposite to 8-oxoG and to a lesser extent opposite to G. hMYH, similar to OGG1, is able to excise 2-OH-A, a mutagenic oxidative adduct. hMYH interacts with several proteins involved in the long-patch repair pathway such as PCNA, Ape1 and RPA.

In addition, hMYH interacts with the mismatch repair proteins hMSH2/hMSH6. Importantly, this interaction enhances the hMYH activity toward A/8-oxo-G. hMYH DNA glycosylase activity is equally enhanced by Ape1 on A/8-oxoG substrates. Unexpectedly this stimulation is independent of Ape1 AP endonuclease activity. Instead, Ape1 acts by stimulation of the formation of hMYH/DNA complexes. These data suggest that different repair pathways may cooperate through protein/protein interactions when counteracting oxidative DNA damage.

Ten different forms of mRNAs coding for hMYH were isolated. These mRNA encodes three proteins of 52, 53 and 57 kDa respectively. The 52 kDa and 53 kDa forms localise in the nucleus and the 57 kDa form in the mitochondria. Expression of hMYH is cell cycle-regulated with a maximum in S phase. Nuclear forms of hMYH co-localise with PCNA at DNA replication foci, suggesting a role of hMYH in replication-associated base excision repair.

Recently biochemical evidence has been provided that human cell extracts perform base excision repair of 8oxo-G/A mismatches on both strands. Similarly to what is reported in yeast, following adenine excision a cytosine is preferentially inserted opposite 8oxoG. This is followed by excision repair of 8oxoG in 8oxoG/C pair.

Interestingly, repair synthesis on either strand is completely inhibited by aphidicolin, suggesting that a replicative DNA polymerase is involved in the gap filling reaction. DNA polymerases/ are likely to be involved in this replication-associated BER. This was confirmed *in vivo* using murine MYH-deficient cells. *MYH-/-* cells repair A/8oxoG mispairs inefficiently but expression of wild-type MYH in these cells increases the repair. Interaction with PCNA is critical, since expression of a functional MYH lacking its PCNA-binding domain had no effect on the repair efficiency of A/8oxoG in *MYH-/-* cells. It has been speculated that hMYH might be considered as a cancer predisposing gene in humans since several mutations in this gene are linked with somatic mutations in *adenomatous polyposis coli* gene (APC).

AP endonucleases Ape1, Ape2: The major human AP endonuclease (HAP-1/Ape1/Apex), homologous to *E. coli* Xth protein (exonuclease III), was independently discovered as an AP-endonuclease and as redox-regulator of the DNA binding domain of Fos-Jun, Jun-Jun, AP-1 proteins and several other transcription factors including NF-kappa B, Myb and members of the ATF/CREB family. Ape1 also activates. Ape1 is a 35.5 kDa nuclear protein, which shows sequence homology to the *E. coli* Xth protein. Targeted homozygous disruption of Apex gene (*Apex-/-*) results in embryonic lethality in mice.

The heterozygous *Apex+/-* mice, so far, did not reveal significant differences in phenotypes associated with oxidative stress as compared to the wild-type. Due to the dual role of Ape1, the early embryonic lethality could be associated either with the BER defect or with the defective regulation of several transcription factors. Ape1 can substitute for exonuclease III in *E. coli* and for Apn1 in *S. cerevisiae*). Beside its AP endonuclease activity, it exhibits other enzymatic activities:

3'5' exonuclease, phosphodiesterase, 3' phosphatase and RNase H. It should be noted that these additional activities are 3-4-fold weaker than its AP endonuclease activity. In addition, Chou and Cheng have shown that Ape1 has a 3'-mismatch exonuclease activity and so it might be considered as a proof-reading enzyme.

Ape1 plays a central role in BER. It is implicated in both short-patch and long-patch repair pathway. It could be also an important regulator of BER. Ape1 interacts with DNA polymerase *in vivo* which is a key element for BER, this interaction permits the loading of pol on DNA at the AP site and the enhancement of its dRPase activity. Ape1 interacts with other BER proteins such as the scaffold protein XRCC1, which stimulates its activity, and with PCNA and the Flap endonuclease 1 (FEN1).

The p21$^{WAF1/CIP1}$ is involved in the regulation of cell cycle progression, DNA replication and DNA repair. It can bind to PCNA inhibiting DNA replication and PCNA stimulation of long-patch BER pathway.

Ape1 seems to be implicated in the compensation of the inhibitory effect of p21 on BER by stimulating and co-ordinating the long-patch BER. In addition, Ape1 is an activator of DNA glycosylases. Altogether these data suggest that BER pathways are co-ordinated and regulated by protein-protein interactions and that Ape1 plays an important role in these interactions. Moreover, it has been shown that Ape1 and hnRNP-L are the components of the nCARE-B2 element binding complexes in the *ape1* gene promoter. This result suggests that expression of the *ape1* gene is down regulated by its own product. Ape1/ref-1 could also be involved in cell response to chemotherapy since it was found that nuclear accumulation of Ape1 is inversely proportional to that of p53 in head-and-neck cancer.

A second AP endonuclease, Ape2, homologous to the *S. cerevisiae* Apn2, was identified in mammals. The *ape2* gene is localised on the X chromosome and Ape2 transcripts are ubiquitously expressed. Ape2, a 57.3 kDa protein, exhibits only a weak ability to complement *E. coli* and *S. cerevisiae* AP endonuclease deficient cells. Ape2 localises predominantly in the nucleus but also in the mitochondria. In the nucleus, Ape2 co-localises with PCNA and interacts with it *in vitro*. These data suggest that Ape2 is implicated in both mitochondrial and nuclear BER and also that the nuclear role of Ape2 is PCNA-dependent.

OTHER ORGANISMS

Evidences have accumulated that oxidative DNA damage are repaired also through BER pathway in plants and insects. Homologues of the *Escherichia coli* Nth and Fpg proteins were characterized in *Arabidopsis thaliana*. Previously it was suggested that *Drosophila* has no BER pathway. However, an homologue of the human OGG1 protein in *Drosophila melanogaster* has been identified and characterized.

A new DNA glycosylase from *A. thaliana* ROS1 is a 1393 aminoacids protein, with an endonuclease III domain at the C-terminus. The recombinant ROS1 is able to incise oxidatively damaged DNA and also DNA containing 5 meC, suggesting a role in controling DNA methylation levels. This agrees with the pleiotropic developmental defects of the ros1 mutants suggesting regulatory properties of this protein. These results suggest that this protein could be a new class of DNA glycosylases, perhaps only present in plants..

ALTERNATIVE REPAIR PATHWAYS

So far, several alternatives to the BER repair pathways for oxidative DNA damage have been described.

Nucleotide Incision Repair

The BER pathway requiring sequential action of two enzymes for incision of DNA raises theoretical problems for the efficient repair of oxidative DNA damage because it generates genotoxic intermediates such as abasic (AP) sites and/or blocking 3'-termini groups that must be eliminated by additional steps before initiating DNA repair synthesis. In addition, biological evidence hints at the existence of an alternative repair pathway. *E. coli* and mouse mutants deficient in the DNA glycosylases that remove oxidised bases are not sensitive to reactive oxygen species.

A recent finding that Nfo and Nfo-like endonucleases nick DNA on the 5' side of various oxidatively damaged bases, generating 3'-hydroxyl and 5'-phosphate termini, provides an alternative pathway to the classic BER. It was proposed to name it the nucleotide incision repair (NIR) pathway. The proposed NIR pathway eliminates the genotoxic intermediates, explains the genetic data and most likely identifies the physiological and original target as the modified base and not an artificial reduced abasic site, for the long patch repair pathway described in human cells.

This alternative repair pathway is evolutionarily conserved from *E. coli* to human. The NIR pathway directly generates proper ends for DNA replication on one side and on the other side the site for elimination of the lesion by specific nucleases. This mechanistic characteristic creates an advantage as compared to BER.

NUCLEOTIDE EXCISION REPAIR

Oxidative DNA damage is a major source of genetic mutation, and it is implicated in several human disorders including *Xeroderma pigmentosum* (XP). Human disorder XP is characterised by defect in nucleotide excision repair, the cells from XP patients fail to remove pyrimidine dimers caused by sunlight and, as a consequence, develop multiple cancers in areas exposed to light.

Despite the fact that neural tissue is shielded from sunlight-induced DNA damage, a fraction of XP patients display progressive neurologic degeneration secondary to a loss of neurones. In fact, it was shown that in *E. coli* the nucleotide excision repair pathway could be involved in repair of oxidative DNA damage. Furthermore, in human cells two major oxidative DNA lesions, 8-oxoguanine and thymine glycol, are excised from DNA *in vitro* by the same enzyme system responsible for removing pyrimidine dimers and other bulky DNA adducts. These results support the idea that the NER pathway could be directly involved in repair of oxidative DNA damage.

Transcription-Coupled Repair (TCR)

Transcription-coupled repair (TCR) is a pathway of DNA excision repair that preferentially removes the DNA lesions present in transcribed sequences of expressed genes. TCR was originally observed for DNA damage repaired by NER pathway. However, recent studies demonstrated that Cockayne syndrome (CS) cells which are deficient for TCR cannot remove 8oxo-G in transcribed sequence, despite its proficient repair in non-transcribed sequence. Furthermore, it has been demonstrated that the breast and ovarian cancer susceptibility genes, BRCA1 and BRCA2, may participate in the repair of the 8oxo-G residues specifically located on the transcribed DNA strand in human cells.

Translesional DNA Synthesis

In *E. coli,* in addition to BER pathway, the translesion DNA synthesis system can contribute to the repair of chromosomal AP sites *in vivo*. In the last few years, new human DNA polymerases Pol-, Pol-, and Pol-, which are able to promote replication through DNA lesions were discovered. These polymerases catalyse translesion DNA synthesis (TLS) in a distributive manner and tend to show lower replication fidelity than other polymerases when unmodified DNA is used as a template. The yeast and human Pol- replicate DNA containing 8-oxo-G efficiently and accurately by inserting a cytosine across from the lesion. In support of the biochemical data, genetic studies show that synergistic increase in the rate of spontaneous mutations occurs in the absence of Pol- in the yeast *ogg1* mutant.

Both error-free and error-prone synthesis were observed in DNA synthesis across A residue catalysed by Pol- and Pol- . It has been found that Pol- preferentially misinsert G opposite to uracil thus providing a mechanism whereby mammalian cells can decrease the mutagenic potential of lesions formed via the deamination of cytosine. Altogether these observations suggest an additional role for TLS-specific DNA polymerases in the prevention of the mutagenic replication of oxidative DNA damage.

DNA Mismatch Repair (MMR) Pathway

DNA mismatch-repair system is involved in the repair of mispaired bases formed during replication, genetic recombination and as a result of DNA damage. Studies in *S. cerevisiae* indicate the involvement of the mismatch repair pathway in prevention of genotoxic effect of oxidative DNA damage. It was shown that mutations in MSH2 and MSH6 caused a synergistic increase in mutation rate in combination with mutations in OGG1, resulting in a 140- to 218-fold increase in the G/C-to-T/A transversion rate. Consistent with this, the MSH2-MSH6 complex binds to 8oxoG:A mispairs and 8oxo-G/C base pairs with high affinity and specificity.

Another important source of 8oxo-G in DNA is the dNTP pool. Recently it has been shown that incorporated 8oxo-dGMP contributes significantly to the mutator phenotype of MMR-deficient cells. Increased expression of MTH1 in MSH2-/- cells produces a significant decrease of DNA 8oxo-G levels and is associated with a drastic reduction of the mutation rate. These findings indicate that MMR excises incorporated 8oxoGMP from newly synthesised DNA strand, thus providing cells with an additional protective strategy from oxidative stress.

Homologous Recombination Repair Pathway

It is well established that homologous recombination (HR) system is pivotal for the repair of an important form of oxidative DNA damage induced by ionising radiation - the double-strand break (DSB). Moreover, *in vivo* replication forks often encounter template DNA damage that can lead to replication fork collapse generating double-strand break. It is hypothesised that DNA recombination functions as a system for reactivation of the collapsed replication fork.

The importance of this function to the formation of lethal DNA damage is underscored by the uncontrollable fragmentation of chromosomes in Rad51-deficient vertebrate cell lines. Another important characteristic of the HR repair pathway is that it mainly provides a non-mutagenic replicative bypass of the blocking lesion using sister chromatid or homologous chromosome as an undamaged DNA template.

Recently, it has been found that defects in the homologous recombination pathway are associated with cancer-prone clinical syndromes, in particular *ataxia telangiectasia,* hereditary breast cancer, Bloom's syndrome and Werner's syndrome. These observations support the idea that the HR system could be an alternative to excision/incision repair pathways in counteraction of genotoxic effects of oxidative DNA damage.

Nucleotide Pool-sanitising Enzymes

Besides the modification of DNA by ROS, these species also oxidise the nucleotide pool. The modified nucleotides could be introduced in DNA and also in RNA during their synthesis leading to mutations. The *E. coli* mutT mutant shows a 100- to 10 000-fold increase in A/TG/C transversion rates as compared to wild-type. This effect is due to the accumulation of mutagenic 8oxo-GTP in the cellular dNTP pool which is incorporated opposite A and C during DNA synthesis, thus leading to A/TG/C transversions. The *E. coli* MuT protein is a 8oxo-GTPase that cleanses the cellular nucleotide pool from 8oxoGTP.

In addition, MutT hydrolyses ribonucleotide 8oxo-GTP, suggesting the importance of this enzyme in maintaining transcription fidelity. Mammalian cells contain MutT homolog, the MTH1 protein (MutT Homolog 1). The corresponding cDNAs were cloned in human and murine cells, and protein was characterised. MTH1 displays only 23% identity to the *E. coli* counterpart but the highly conserved region essential for the 8oxo-GTPase activity contains identical residues.

Therefore, expression of MTH1 human and rodent origins in *E. coli mutT* reverts to the mutator phenotype. These results indicate that mammalian proteins may have a similar antimutagenic activity *in vivo* as *E. coli* MutT protein. Like MutT, MTH1 hydrolyses 8oxo-GTP and oxidised ribonucleotides 8oxo-GTP. Interestingly, MTH1 possess an additional 2-hydroxyadenosine (2-OH-A) triphosphate pyrophosphorylase activity. Since hMYH (MutY homolog) is able to remove 2-OH-A from DNA, these data suggest that mammalian cells possess a pathway to sanitise oxidised forms of both guanine and adenine.

Seven types of mRNA transcripts, coded by the MTH1 gene, have been identified in human Jurkat cells, which lead to the synthesis of, at least, three polypeptides of 18, 21 and 22 kDa. MTH1 polypeptides localise mainly in the cytoplasm and in mitochondria.

The MTH1 knockout cell lines and mice were constructed by gene targeting. MTH1-/- cells exhibit a twofold increase of mutation rate as compared to MTH1+/+ cells which is in contrast to the increase of mutation rate in *E. coli mutM* cells. Therefore, it seems likely that mammalian cells have other enzyme(s) and/or pathway(s) capable of hydrolysing oxidised nucleotides. MTH1-/- mice exhibit slightly elevated tumour incidence than wild-type mice but their survival at 1.5 years is not affected.

Several studies on human cancers show that mutations in hMTH1 gene, so far are not involved in these pathologies. Unexpectedly, MTH1 and translesional DNA polymerase kappa (Pol-) expressions were found to significantly increase and decrease respectively in two mammary carcinoma cell lines characterised by frequent A/TG/C transversions. So far, no human diseases have been linked to defects in proteins involved in the BER pathway. DNA repair genes functionally expressed in mammalian cells and now in transgenic mice having a null mutation in the gene coding for BER proteins are very important tools to ascertain the biological role of these proteins *in vivo*.

It has been very astonishing to notice that, beside a few examples of targeted deletion of genes encoding the AP-endonuclease 1 and DNA polymerase in mice leading to embryonic lethality, the genotype of the other BER protein knockout mice (such as a number of DNA glycosylases) does not show any striking particularity in terms of predisposition to cancer and premature aging.

Furthermore, examination of 6200 *S. cerevisiae* genes transcript levels after exposure to various genotoxic agents reveals that the DNA repair genes are only modestly induced, which is in contrast to adaptive and SOS response in *E. coli*. These observations suggest the possibility of back-up repair pathway(s) for oxidative DNA damage that have to be characterized. One could expect important breakthroughs from crosses between different strains to produce double knockouts to identify the possible back-up systems, the processes involved in the regulation and the interactions of the different pathways. Detailed understanding of the mechanisms leading to the co-ordination of various proteins involved in the molecular reaction of BER is of paramount importance for gaining insights into the efficiency and fidelity of this key pathway for genome stability, prevention of cancer, resistance to chemotherapeutic agents, degenerative diseases and more recently in some aspects of teratogenicity.

Chapter 6

Transcription Factors

Transcription factors play essential role in the gene regulation in higher organisms, binding to multiple target sequences and regulating multiple genes in a complex manner. In order to understand the molecular mechanism of target recognition, and to predict target genes for transcription factors at the genome level, it is important to analyse the relationship between the structure and function (specificity) of transcription factors. We have used a knowledge-based approach, utilizing rapidly increasing structural data of protein-DNA complexes, to derive empirical potential functions for the specific interactions between bases and amino acids as well as for DNA conformation, from the statistical analyses on the structural data.

Then these statistical potentials are used to quantify the specificity of protein-DNA recognition. The quantification of specificity has enabled us to establish the structure-function analysis of transcription factors, such as the effects of binding cooperativity on target recognition. The method is also applied to real genome sequences, predicting potential target sites. We are also using computer simulations of protein-DNA interactions in order to complement the empirical method. Combining the two approaches together, we can better understand the mechanism of protein-DNA recognition and improve the target prediction of transcription factors.

Regulation of gene expression in higher organisms is achieved by a specific recognition of target DNA sequences by DNA-binding proteins. Due to the progress of X-ray crystallography and NMR spectroscopy techniques, structural data on the protein-DNA complexes have been rapidly increasing. However, the mechanism of DNA sequence recognition by proteins has been poorly understood, and thus the accurate prediction of their targets at the genome level is not yet possible.

This situation implies that the structural information has not been fully utilized. Understanding the molecular mechanism and its application to genomewide prediction are essential for the analysis of gene regulation network. Here, we describe two kinds of approaches for studying protein-DNA recognition. One is a knowledge-based approach, by which we extract functional information from structural data of protein- DNA complexes. We made a statistical analysis of structural database of protein-DNA complex, and derived empirical potential functions for the specific interactions between bases and amino acids.

Then, we used a sequence-structure threading to examine the relationship between structure and specificity in protein-DNA recognition, and to predict target sites of transcription factors in real genome sequences. We also evaluated the fitness of DNA sequence against DNA structure to examine the role of indirect readout mechanism. We show how the structural features are related to the specificity, and discuss relative roles of direct and indirect readout mechanisms in the recognition. We also discuss the strategy to predict the target sequences of transcription factors at the genome level, and its application to yeast genome.

In another approach, we analyse protein-DNA recognition by computer simulations. We describe several different methods of computer simulations. We extracted interacting pairs of bases and amino acids from a set of non-redundant protein-DNA complex structures. In order to derive the statistical potential of interactions between bases and amino acids, we analyzed the amino acid distributions around each base. We defined a coordinate system by taking an origin N9 atom for A and G and N1 atom for T and C.

We considered the amino acids within a given box, and the box was divided into grids. Because the sample numbers are not yet very large, we first consider the information about Ca atoms. Then we transformed the distributions of Ca atom of amino acid into statistical potentials defined by the following equations,

$$\Delta E^{ab}(s) = -RT \ln \frac{f^{ab}(s)}{f(s)}$$

$$f^{ab}(s) = \frac{1}{1+m_{ab}w} f(s) + \frac{m_{ab}w}{1+m_{ab}w} g^{ab}(s)$$

where *m*ab is the number of pairs, amino acid *a* and base *b* observed, *w* is the weight given to each observation, *f*(*s*) is the relative frequency of occurrence of any amino acids at grid point *s*, and $g^{ab}(s)$ is the equivalent relative frequency of occurrence of amino acid *a* against base *b*. *R* and *T* are gas constant and absolute temperature, respectively. Here, we used a box of |x| = |y| =13.5Å and |z| = 6 Å and a grid interval of 3Å, which was determined by examining various intervals. By threading a set of random DNA sequences onto the template structure, we calculated the Z-score of the specific sequences against the random sequences, which represents the specificity of the complex. Assuming the additivity of potential energies, the sum of the potential energies ($E_{PD} = \Sigma ab,s\ \Delta Eab(s)$) for a given DNA sequence in a complexed form was defined as the energy for the sequences. The energy for a particular sequence, in a crystal structure for example, was normalized to measure specificity by the Z-score against random sequences.

The Z-score was defined as (X - *m*)/s, where *X* is the energy of a particular sequence, *m* is the mean energy of 50,000 random DNA sequences, and s is the standard deviation. We have also derived statistical potential functions for conformational energy of DNA from the protein-DNA complex structural data to evaluate the fitness of sequences to a particular conformation of DNA. To estimate the sequence-dependent DNA conformational energy, we mostly followed the approach described by Olson et al.

The conformation energies were approximated using a harmonic function, $E_{DNA} = 1/2\ \Sigma\Sigma\ f_{ij}\ \Delta\theta_i\ \Delta\theta_j$, in which qi represents the base-step parameters, and *fij* are the elastic force constants impeding deformation of the given base step and $\Delta\theta_i = \theta_i - \theta_i^0$, in which qi^0 is the average base-step parameter. The basestep parameters used were shift, slide, rise, tilt, roll, and twist. The unknown parameters f_i^j and θ_i^0 were determined by statistical analysis of the same nonredundant protein-DNA complexes.

Setting up a covariance matrix from observed distributions of θ_i thus refers to an effective inverse harmonic force-constant matrix. Inversion of this matrix transformed it to a forceconstant matrix in the original coordinate basis. We removed all parameters of a base step for which one parameter exceeded three standard deviations, in a iterative manner. Then the final force field was calculated. The conformational energy of DNA in a given complex structure was calculated as the sum of all the base steps.

We assigned the energy corresponding to the threshold value when any parameter exceeded three standard deviations. Then, thesc potentials were used to quantify the specificity of indirect readout mechanism of protein-DNA recognition, as a Z-score, by using the same threading procedure. We carried out jack-knife and bootstrap tests to assess the statistical confidence in the Z-score calculations. To remove the effect of self contributions, we always removed the self from the original dataset of complex structures when calculating its Z-score.

We then examined the Z-score further by removing one additional randomly selected structure from the dataset and repeated this procedure. We found that the Z-scores were stable against these treatments, indicating that our dataset of protein-DNA complexes provides adequate information for a generally valid structure-based potential. To calculate the bootstrap standard errors, we prepared a set of 61 randomly selected complex structures in which the self was removed but duplications of the same structure were allowed.

We created 200 such replications and calculated the standard errors. We combined the direct and indirect energies to calculate the total energy. Because the derivations of these empirical energies are based on different statistics, we cannot simply make a summation.

We need to introduce a weighting factor; Etot = cEPD + (1-c)EDNA, where EPD and EDNA are the energies of the direct and indirect readout, respectively, and c is a weighting coefficient ranging between 0 and 1. This coefficient is determined by maximizing the total Z-score – i.e., the Zscore is calculated from random sequences, and a value of c is sought that gives the highest total Z-score. Then, the threading procedure was applied to the real genome sequences in order to find potential target sites, by using the total energy.

We carry out several different types of computer simulations depending on the resolution of the system under investigation: Monte Carlo simulation of base amino acid interactions; molecular dynamics and free energy calculations of protein-DNA complex; empirical calculations of interaction energy of protein-DNA complex: and docking simulation of protein-DNA binding and sliding.

RELATIONSHIP IN PROTEIN-DNA RECOGNITION

It is interesting to compare two structures, cognate and non-cognate complex structures in order to understand what is important for specific binding and what is different between them. There are several such examples in PDB (Protein Data Bank). We examined NF-kB, glucocorticoid receptor DNA binding domain, EcoRV endonuclease and BamHI endonuclease, for which both the cognate and non-cognate complex structures are available in PDB.

The statistical method could distinguish the two structures as differences in the Zscores as well as statistical poteitials. Thus, the subtle differences in specificity of these structures could be detected by the analysis of energies. Proteins often bind to DNA as homodimers, which leads to subtle structural differences between the two subunits. Thus, we examined in detail the structural effects of asymmetric binding on specificity.

Marked differences in the specificity of DNA binding were observed for the two subunits of l repressor, the glucocorticoid receptor, and for transcription factors containing a Zn_2CyS_6 binuclear cluster domain, which are known to bind asymmetrically to DNA. Table shows some examples of asymmetric Z-scores for homodimeric protein-DNA complexes.

Table: Z-scores for Homodimeric Protein-DNA Complexes. Z-scores were Calculated for the whole Complex and Two Monomers Separately.

Protein-DNA Complex	*Z-scores*		
	Whole	*Chain 1*	*Chain 2*
Pyrimidine pathway regulator	–2.9	–3.1	–1.1
λ repressor	–2.6	–2.9	–0.1
Glucocorticoid Receptor	–1.1	–1.8	0.6

We also applied this method to examine the relationship between structure and specificity in cooperative protein-DNA binding. The effect of cooperative binding was examined by comparing the monomer and heterodimer complexes of MATa1/a2, MCM1/MATa2 and NFAT/AP- 1 transcription factors. We found that the heterodimer binding enhances the specificity in a non-additive manner.

This result indicates that the conformational changes introduced by the heterodimer binding play an important role in enhancing the specificity. The structures of cognate and noncognate complexes of EcoRV show marked differences in the conformations of both the enzyme and DNA. The example of EcoRV enabled us to show the cooperative effect of sequence and structure on specificity - i.e., conformational differences between the cognate and noncognate complexes were sensitive to the cognate but not the noncognate DNA sequence, and the cognate structure had a greater ability to discriminate DNA sequences than the noncognate structure. The above method is based on the direct readout mechanism, in which protein recognizes DNA sequence through the direct contact between amino acids and base pairs.

On the other hand, substitutions of those base pairs not in contact with amino acids often affect binding affinity, indicating that protein may recognize DNA sequence through particular structure or property of DNA. This indirect readout mechanism may contribute to the specificity of protein-DNA recognition significantly. We have derived statistical potential functions for conformational energy of DNA to quantify the specificity of indirect readout mechanism of protein- DNA recognition. Once the potentials are derived, the conformational energy of DNA sequence can be estimated for given structure, and the threading procedure can be used to evaluate the fitness of sequence to structure of DNA.

We can calculate the Z-score for the indirect recognition in the same way as for the direct recognition. By comparing both the Z-scores we can assess the relative contributions of direct and indirect readout mechanisms. We have analyzed various protein- DNA complexes systematically, and found that both the direct and indirect mechanisms make significant contribution to the specificity. The relative contributions depend on the types of DNAbinding proteins.

Because both the potentials are independent quantities, they can be summed up to calculate the total energy and used to find target sites, although a weighting factor needs to be determined as the two potentials were derived from different statistics. The Zscore calculated for the total energy with varying weighting coefficient. One can see enhanced Z-score compared with individual Z-scores for direct or indirect readout (c=1 or 0, respectively).

This result indicates that the energies of the direct and indirect readouts contain independent information that in combination enhances the specificity of the recognition. We use the weight coefficient that gives the maximum total Z-score. We use the combined energy for threading against DNA sequences to make target predictions for transcription factors.

Prediction in Yeast Genome

The threading procedure was used to find target sites of transcription factors in real genome sequences. As an example of such applications, we could identify the experimentally-verified binding sites of the transcription factor MATa1/a2 in the promoter of *HO* gene successfully. We have also attempted to identify target sites and genes of MCM1/MATa2 in the whole yeast genome. The target genes of this transcription factor have been identified in yeast genome experimentally. The predicted target genes were ranked by the Z-score and compared with experimental data.

The target genes identified positively by experiment were ranked high in the list, and the experimentally negative genes were ranked low. Separation between the positive and negative genes was not perfect but they were segregated by a certain threshold Z-score value. The total Z-score gave better separation than that of direct contribution alone. We are applying the method for the targets prediction of other transcription factors.

PROTEIN-DNA INTERACTIONS

The statistical method has shown that the distribution of Ca position of amino acids around base pairs provides important information about the specificity in the DNA sequence recognition by proteins. However, the accuracy of this prediction method is limited by the number of available structural data.

Thus, it is desirable to complement the method by some other means. Computer simulation is one such method. In real DNA-protein interactions, however, there are many factors contributing to the recognition process. Thus, it is necessary to examine different levels of the system by different methods.

FREE-ENERGY MAP FOR BASE-AMINO ACID INTERACTIONS

In the case of base-amino acid interactions, we need to reproduce the distribution of Ca position of amino acids around base pairs observed in the experimental DNAprotein complex structures. Thus, it would be reasonable to consider at first the interactions between a base pair and an amino acid. In reality, the Ca position is fixed by main chain of protein and the possible range of Cb direction may be restricted. However, such biases, which are context dependent, are difficult to evaluate *a priori*.

Therefore, at first we will consider intrinsic interactions between a base pairs and an amino acid. We generated many Ca positions and side chain conformations by systematic sampling or Monte Carlo sampling methods, and calculated free energy map of Ca around a given base pair. By calculating the free energies for different Ca positions and subtracting a reference free energy at a large separation, we can obtain a contour map of interaction free energy, which shows preferable positions of Ca of amino acid around a base pair. This can be directly compared with the distribution of Ca position of amino acids around base pairs derived from DNA-protein complex database.

According to the freeenergy contour maps for the interactions of Asn with AT and with G-C, the preferable position of Ca is localized in a narrow region around A in the case of A-T. In this region, Asn and A form specific double hydrogen bonds, C=O...HN6 and NH...N7, which are found frequently in the Asn-A pair in the experimental structures of DNA-protein complexes. Also, the distribution of Ca is in agreement with the statistical potential obtained by the database analysis.

On the other hand, Asn tends to be more broadly distributed around G-C. The lowest DG values are located in the middle of G-C, and the following lower *DG's* extend towards C5 atom of C, where C does not have a methyl group. This comparison indicates that the interaction of Asn is more specific toward A-T than G-C. This example illustrates how we can quantify the specificity in the base-amino acid interactions and complement the statistical method.

SEMI-EMPIRICAL ESTIMATION OF FREE-ENERGY CHANGES DUE TO BASE MUTATIONS

The specificity of protein-DNA recognition is usually assessed by the effect of mutations. If mutations cause large effect on the binding affinity, it would indicate that specific interactions are involved at the mutation site, and the extent of specificity dictates the effect of mutation. Thus, it is important to examine the effect of mutation and component of interaction energy causing the effect, as well as to develop methods to predict the mutation effect. We have performed a computer analysis in which a phage DNA-binding protein, l repressor, was used to examine the changes in binding free-energy (DDG) and its energy components caused by single base mutations. We then determined which of the calculated energy components best correlated with the experimental data. The experimental DDG values were well reproduced by the calculations.

Component-analysis revealed that the electrostatic and hydrogen bond energies were most strongly correlated with the experimental data. Among the 51 single basesubstitution mutants examined, positive DDG values, corresponding to weakened binding, were caused by the loss of favorable electrostatic interactions and hydrogen bonds, the introduction of steric collisions and electrostatic repulsion, the loss of favorable interactions with a thymine methyl group, and the increase of unfavorable hydration energy from isolated DNA.

These results suggest that electrostatic interactions and hydrogen bond make major contributions to the specificity of l repressor-DNA recognition. However, van der Waals and hydrophobic interactions also play important role in particular DNA sequence locations.

FREE-ENERGY CALCULATIONS OF PROTEIN-DNA COMPLEXES

For a large system like l repressor DNA complex, it is not an easy task to use all-atom simulations to calculate interaction free energy changes caused by mutations accurately. However, calculation of free energy changes due to small structural perturbation in the complex is feasible. We calculated a binding free energy change between DNA and l repressor due to a base substitution from tymine (T) to uracil (U) by the free energy perturbation method based on molecular dynamics simulations for the complex in water with all degrees of freedom and long-range Coulomb interactions.

The binding free energy change calculated was in good agreement with an experimental value. By using component analysis, we could clarify the reason why the small difference between T and U (CH3 in T is replaced by H in U) caused the significant binding free energy change. The free energy change is due to the gain of hydration free energy in the dissociated state and the loss of favorable van der Waals interactions in the associated state.

Docking Simulations of Protein-DNA Complexes

During the recognition process, proteins associate and dissociate with DNA, and may slide along DNA before finding their target sites. Thus, protein-DNA recognition is a dynamical process. In order to reveal reaction pathways of proteins along DNA, we are conducting docking simulations of protein-DNA complexes by using Monte Carlo method. At first, using rigid-body approximations, we examined free-energy profile along the DNA surface for homeodomains and restriction enzymes.

Preliminary results indicate that the homeodomain may track along the major groove of DNA while it slides on DNA. We will incorporate the flexibilities of protein and DNA molecules such as hinge and bending motions, and examine the role of these effects on the protein-DNA recognition process.

We have described two kinds of methods for studying the relationship between structure and function (specificity) in protein-DNA recognition. One is the knowledge-based approach, by which we extract biological information from structural data on protein- DNA complexes. The increase in the structural data, which will be accelerated by structural genomics project, will make this structure-based method promising for revealing the structure-function relationship in protein- DNA recognition and for predicting targets of transcription factors.

This method can also be applied to proteins of unknown structure having substantial sequence similarity to known proteins: the structure can be modeled based on the similarity and its binding sites can be predicted. This approach is, however, limited by the number of available data. Thus, we try to complement it with a deductive approach, by which we try to reproduce the recognition process by computer simulation. Although this approach requires a large amount of computational time, the advancement of technology makes the computer simulation of large systems feasible.

The information obtained by this approach would help the knowledge-based approach in improving the accuracy of statistical potentials and target prediction. Thus, a combination of the two methods will become a powerful tool for the study of protein-DNA recognition and its application to target prediction at the genome level.

Chapter 7

Gene Expression

GENE REGULATION

Gene Regulation refers to the cellular control of the amount and timing of changes to the appearance of the functional product of a gene. Although a functional gene product may be an RNA or a protein, the majority of known mechanisms regulate protein coding genes. Any step of the gene's expression may be modulated, from DNA-RNA transcription to the post-translational modification of a protein.

Gene regulation is essential for viruses, prokaryotes and eukaryotes as it increases the versatility and adaptability of an organism by allowing the cell to express protein when needed. The first example of gene regulation system was the lac operon, discovered by Jacques Monod, in which protein involved in lactose metabolism are expressed by E.coli only in the presence of lactose and absence of glucose.

Furthermore, gene regulation allows the presence in a multicellular organism of different cells types arranged in a complex pattern, hence different transcriptomes despite them having all the same genome and the generation of patterns by cellular differentiation and morphogenesis.

REGULATED STAGES OF GENE EXPRESSION

Any step of gene expression may be modulated, from the DNA-RNA transcription step to post-translational modification of a protein. The following is a list of stages where gene expression is regulated:

- Chromatin domains

- Transcription
- Post-transcriptional modification
- RNA transport
- Translation
- mRNA degradation
- Post-translational modifications

MODIFICATION OF DNA

In eukaryotes, the accessibility of large regions of DNA can depend on its chromatin structure which can be altered as a result of histone modifications which are directed by DNA methylation, ncRNA or DNA binding protein.

Chemical

Methylation of DNA is a common method of gene silencing. DNA is typically methylated by methyltransferase enzymes on cytosine nucleotides in a CpG dinucleotide sequence (also called "CpG islands" when densely clustered). Analysis of the pattern of methylation in a given region of DNA (which can be a promoter) can be achieved through a method called bisulfite mapping.

Methylated cytosine residues are unchanged by the treatment, whereas unmethylated ones are changed to uracil. The differences are analyzed by DNA sequencing or by methods developed to quantify SNPs, such as Pyrosequencing (Biotage) or MassArray (Sequenom), measuring the relative amounts of C/T at the CG dinucleotide. Abnormal methylation patterns are thought to be involved in carcinogenesis.

Structural

Transcription of DNA is dictated by its structure. In general, the density of its packing is indicative of the frequency of transcription. Octameric protein complexes called histones are responsible for the amount of supercoiling of DNA, and these complexes can be temporarily modified by processes such as phosphorylation or more permanently modified by processes such as methylation.

Such modifications are considered to be responsible for

more or less permanent changes in gene expression levels. Histone acetylation is also an important process in transcription.

Histone acetyltra-nsferase enzymes (HATs) such as CREB-binding protein also dissociate the DNA from the histone complex, allowing transcription to proceed.

Often, DNA methylation and histone deacetylation work together in gene silencing. The combination of the two seems to be a signal for DNA to be packed more densely, lowering gene expression.

REGULATION OF TRANSCRIPTION

Regulation of transcription controls when transcription occurs and how much RNA is created. Transcription of a gene by RNA polymerase can be regulated by at least five mechanisms:

- *Specificity factors* alter the specificity of RNA polymerase for a given promoter or set of promoters, making it more or less likely to bind to them (i.e. sigma factors used in prokaryotic transcription).
- *Repressors* bind to non-coding sequences on the DNA strand that are close to or overlapping the promoter region, impeding RNA polymerase's progress along the strand, thus impeding the expression of the gene.
- *General transcription factors* These transcription factors position RNA polymerase at the start of a protein-coding sequence and then release the polymerase to transcribe the mRNA.
- *Activators* enhance the interaction between RNA polymerase and a particular promoter, encouraging the expression of the gene. Activators do this by increasing the attraction of RNA polymerase for the promoter, through interactions with subunits of the RNA polymerase or indirectly by changing the structure of the DNA.
- *Enhancers* are sites on the DNA helix that are bound to by activators in order to loop the DNA bringing a specific promoter to the initiation complex.

POSTTRANSCRIPTIONAL REGULATION

After the DNA is transcribed and mRNA is formed there must be some sort of regulation on how much the mRNA is translated into Proteins. Cells do this by modulating the Capping, Splicing, addition of a Poly(A) Tail, the sequence-specific nuclear export rates and in several contexts sequestration of the RNA transcript. These processes occur in eukaryotes but not in prokaryotes. This modulation is a result of a protein or transcript which in turn is regulated and may have an affinity for certain sequences.

- *Capping* changes the five prime end of the mRNA to a three prime end by 5'-5' linkage, which protects the mRNA from 5' exonuclease, which degrades foreign RNA. The cap also helps in ribosomal binding.
- *Splicing* removes the introns, noncoding regions that are transcribed into RNA, in order to make the mRNA able to create proteins. Cells do this by spliceosome's binding on either side of an intron, looping the intron into a circle and then cleaving it off. The two ends of the exons are then joined together.
- *Addition of poly(A) tail* otherwise known as poly-adenylation. Junk RNA is added to the 3' end, and acts as a buffer to the 3' exonuclease in order to increase the half life of mRNA.

In both prokaryotes and eukaryotes a large number of RNA binding protein exist, with often are directed to their target sequence by the secondary structure of the transcript, which may change depending on certain conditions, such as temperature or presence of a ligand (aptamer), some transcripts act as ribozymes and self-regulate their expression.

EXAMPLES OF GENE REGULATION

- Enzyme induction is a process in which a molecule (e.g. a drug) induces (i.e. initiates or enhances) the expression of an enzyme.
- The induction of heat shock proteins in the fruit fly *Drosophila melanogaster*.

- The Lac operon is an interesting example of how gene expression can be regulated.
- Viruses despite having only a few genes, possess mechanisms to regulate their gene expression, typically into a early and late phase, using collinear systems regulated by anti-terminators (lambda phage) or splicing modulators (HIV)

CIRCUITRY

UP-REGULATION AND DOWN-REGULATION

Up-regulation is a process which occurs within a cell triggered by a signal (originating internal or external to the cell) which results in increased expression of one or more genes and as a result the protein(s) encoded by those genes. Conversely down-regulation is a process resulting in decreased gene and corresponding protein expression.

Up: Regulation occurs for example when a cell is deficient in some kind of receptor. In this case, more receptor protein is synthesized and transported to the membrane of the cell and thus the sensitivity of the cell is brought back to normal reestablishing homeostasis.

Down: Regulation occurs for example when a cell is overly stimulated by a neurotransmitter, hormone, or drug for a prolonged period of time and the expression of the receptor protein is decreased in order to protect the cell.

INDUCIBLE Vs. REPRESSIBLE SYSTEMS

Gene Regulation can be summarized as how they respond:

Inducible systems: An inducible system is off unless there is the presence of some molecule (called an inducer) that allows for gene expression. The molecule is said to "induce expression". The manner in which this happens is dependent on the control mechanisms as well as differences between prokaryotic and eukaryotic cells.

Repressible Systems: A repressible system is on except in the presence of some molecule (called a corepressor) that suppresses gene expression. The molecule is said to "repress expression". The manner in which this happens is dependent on the control mechanisms as well as differences between prokaryotic and eukaryotic cells.

DEVELOPMENTAL BIOLOGY

A large number of studied regulatory systems come from developmental biology. Examples include:

The collinearity of the Hox gene cluster with their nested antero-posterior patterning.

It has been speculated that pattern generation of the hand (digits - interdigits) The gradient of Sonic hedgehog (secreted inducing factor) from the zone of polarizing activity in the limb which creates a gradient of active Gli3 which activates Gremlin which inhibits BMPs also secreted in the limb resulting in the formation of an alternating pattern of activity as a result of this reaction-diffusion system.

Somatogenesis is the creation of segmatation (somites) form a uniform tissue (PSM) sequentially from anterior to posterior, this is a achieved in amniotes possibly by means of two opposing gradients, Retinoic acid in the anterior (wavefront) and an oscillating gradient in the posterior (clock) composed of FGF + Notch and Wnt in antiphase. Sex determination in the soma of a Drosophila requires the sensing of the ratio of autosomal genes to sex chromosome encoded genes, which results in the production of sexless splicing factor in females resulting in the female isoform of doublesex.

THEORETICAL CIRCUITS

- Repressor/Inducer: a activation of a sensor results in the change of expression of a gene
- Negative feedback: the gene product downregulates its own production directly or indirectly, which can result in
  - Keeping transcript levels constant/proportional to a factor
  - Inhibition of run-away reactions when coupled with a positive feedback loop
  - Creating an oscillator by taking advantage in the

time delay of transcription and translation, given that the mRNA and protein half-life is shorter

- Positive feedback: the gene product upregulates its own production directly or indirectly, which can result in
  - Signal amplification
  - Bistable switches when two genes inhibit each other and have both positive feedback
  - Pattern generation

Methods

Generally, most experiments investigating differential expression used whole cell extracts of RNA, called steady-state levels, to determined which genes changed and by how much they did, these are however not informative of where the regulation has occurred and may actually mask conflicting regulatory processess, it is the most commonly analysed (QPCR and DNA microarray).

When studying gene expression there are several methods to look at the various stages. In eukaryotes these include:

The chromatin conformation of the region can be determined by ChIP-chip analysis by pulling down RNA Polymerase II, Histone 3 modifications, Trithorax-group protein,Polycomb-group protein or any other DNA binding element to which a good antibody is available.

Due to post-transcriptional regulation, transcription rates and total RNA levels differ significantly, to measure the transcription rates nuclear run-on assays can be done and newer high-throughput methods are being developed, using thiol labelling instead of radioactivity.

Only 5% of the RNA polymerised in the nucleus actually exists and not only introns, abortive products and non-sense transcripts are degradated therefore the differences in nuclear and cytoplasmic levels can be see by separating the two fractions by gentle lysis.

Alternative splicing can be analysed with a splicing array or with a tiling array. All in vivo RNA is complexed as RNPs. The quantity of transcripts bound to specific protein can be also analysed by RIP-Chip, for example DCP2 will give an indication of sequestered protein, ribosome bound gives and indication of transcripts active in transcription (although it should be noted that a more dated method, called polysome fractionation, is still popular in some labs)

Protein levels can be analysed by Mass spectrometry, which can only be compare to QPCR data as microarray data is relative and not absolute. RNA and protein degradation rates are measured by means of transcription inhibitors or translation inhibitors (Cycloheximide) respectively.

GENE EXPRESSION

Gene expression is the process by which inheritable information from a gene, such as the DNA sequence, is made into a functional gene product, such as protein or RNA. Several steps in the gene expression process may be modulated, including the transcription step and the post-translational modification of a protein. Gene regulation gives the cell control over structure and function, and is the basis for cellular differentiation, morphogenesis and the versatility and adaptability of any organism.

Gene regulation may also serve as a substrate for evolutionary change, since control of the timing, location, and amount of gene expression can have a profound effect on the functions (actions) of the gene in the organism. Non-protein coding genes (e.g. rRNA genes, tRNA genes) are transcribed, but not translated into protein.

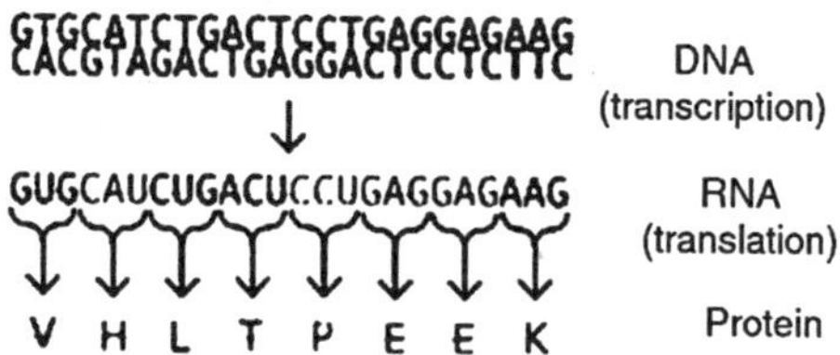

Genes are expressed by being transcribed into RNA, and this transcript may then be translated into protein.

MEASUREMENT

The expression of many genes is regulated after

transcription (i.e., by microRNAs or ubiquitin ligases), so an increase in mRNA concentration need not always increase expression. Nevertheless, mRNA levels can be quantitatively measured by Northern blotting, a process in which a sample of RNA is separated on an agarose gel and hybridized to a radio-labeled RNA probe that is complementary to the target sequence. Northern blotting requires the use of radioactive reagents and can have lower data quality than more modern methods (due to the fact that quantification is done by measuring band strength in an image of a gel), but it is still often used. It does, for example, offer the benefit of allowing the discrimination of alternately spliced transcripts.

A more modern low-throughput approach for measuring mRNA abundance is real-time polymerase chain reaction (The term RT-PCR is used to refer to both reverse transcription PCR as well as real-time PCR, which is also known as quantitative RT-PCR or quantitative PCR (qPCR). With a carefully constructed standard curve qPCR can produce an absolute measurement such as number of copies of mRNA per nanolitre of homogenized tissue. The lower level of noise in data obtained via qPCR often makes this the method of choice, but the price of the required equipment and reagents can be prohibitive.

In addition to low-throughput methods, transcript levels for many genes at once (expression profiling) can be measured with DNA microarray technology or "tag based" technologies like Serial analysis of gene expression (SAGE) or the more advanced version SuperSAGE, which can provide a relative measure of the cellular concentration of different messenger RNAs. Recent advances in microarray technology allow for the quantification, on a single array, of transcript levels for every known gene in the human genome.

The great advantage of tag-based methods is the "open architecture", allowing for the exact measurement of any transcript, known or unknown. Especially SuperSAGE recommends itself therefore also for studying organisms with unknown genomes.

Protein levels themselves can be estimated by a number

of means. The most commonly used method is to perform a Western blot against the protein of interest, whereby cellular lysate is separated on a polyacrylamide gel and then probed with an antibody to the protein of interest. The antibody can either be conjugated to a fluorophore or to horseradish peroxidase for imaging or quantification. Another commonly used method for assaying the amount of a particular protein in a cell is to fuse a copy of the protein to a reporter gene such as Green fluorescent protein, which can be directly imaged using a fluorescent microscope.

Because it is very difficult to clone a GFP-fused protein into its native location in the genome, however, this method often cannot be used to measure endogenous regulatory mechanisms (GFP-fusions are therefore most often expressed on extra-genomic DNA such as an expression vector). Fusing a target protein to a reporter can also change the protein's behaviour, including its cellular localization and expression level. The pattern of detection of a gene or gene product may be described using terms such as facultative, constitutive, circadian, cyclic, housekeeping, or inducible.

REGULATION OF GENE EXPRESSION

Regulation of gene expression is the cellular control of the amount and timing of appearance of the functional product of a gene. Any step of gene expression may be modulated, from the DNA-RNA transcription step to post-translational modification of a protein. Gene regulation gives the cell control over structure and function, and is the basis for cellular differentiation, morphogenesis and the versatility and adaptability of any organism.

Expression System

An expression system consists, minimally, of a source of DNA and the molecular machinery required to transcribe the DNA into mRNA and translate the mRNA into protein using the nutrients and fuel provided. In the broadest sense, this includes every living cell capable of producing protein from DNA. However, an expression system more specifically refers to a laboratory tool, often artificial in some manner, used for assembling the product of a specific gene or genes. It is defined as the "combination of an expression vector, its cloned DNA, and the host for the vector that provide a context to allow foreign gene function in a host cell, that is, produce proteins at a high level".

In addition to these biological tools, certain naturally observed configurations of DNA (genes, promoters, enhancers, repressors) and the associated machinery itself are referred to as an expression system, as in the simple repressor 'switch' expression system in Lambda phage. It is these natural expression systems that inspire artificial expression systems, (such as the Tet-on and Tet-off expression systems).

Each expression system has distinct advantages and liabilities, and may be named after the host, the DNA source or the delivery mechanism for the genetic material. For example, common expression systems include bacteria (such as E.coli), yeast (such as S.cerevisiae), plasmid, artificial chromosomes, phage (such as lambda), cell lines, or virus (such as baculovirus, retrovirus, adenovirus).

Over Expression

In the laboratory, the protein encoded by a gene is sometimes expressed in increased quantity. This can come about by increasing the number of copies of the gene or increasing the binding strength of the promoter region.

Often, the DNA sequence for a protein of interest will be cloned or subcloned into a plasmid containing the *lac* promoter, which is then transformed into the bacterium *Escherichia coli*. Addition of IPTG (a lactose analog) causes the bacteria to express the protein of interest. However, this strategy does not always yield functional protein, in which case, other organisms or tissue cultures may be more effective.

For example, the yeast *Saccharomyces cerevisiae* is often preferred to bacteria for proteins that undergo extensive posttranslational modification. Nonetheless, bacterial expression has the advantage of easily producing large amounts of protein, which is required for X-ray crystallography or nuclear magnetic resonance experiments for structure determination.

GENE NETWORKS AND EXPRESSION

Genes have sometimes been regarded as nodes in a network, with inputs being proteins such as transcription factors, and outputs being the level of gene expression. The node itself performs a function, and the operation of these functions have been interpreted as performing a kind of information processing within cell and determine cellular behaviour.

CONTROL OF GENE EXPRESSION

While the period from 1900 to the second world war has been called the "golden age of genetics", we may be in a new golden (or platinum) age. Recombinant DNA technology allows us to manipulate the very DNA of living organisms and to make conscious changes in that DNA. Prokaryote genetic systems are much easier to study and better understood than are eukaryote systems.

The Chromosome of *E. Coli*

The single chromosome of the common intestinal bacterium *E. coli* is circular and contains some 4.7 million base pairs. It is nearly 1 mm long, but only 2nm wide. The chromosome replicates in a bidirectional method, producing a figure resembling the Greek letter theta. The promoter is the part of the DNA to which the RNA polymerase binds before opening the segment of the DNA to be transcribed.

A segment of the DNA that codes for a specific polypeptide is known as a structural gene. These often occur together on a bacterial chromosome. The location of the polypeptides, which may be enzymes involved in a biochemical pathway, for example, allows for quick, efficient transcription of the mRNAs. Often leader and trailer sequences, which are not translated, occur at the beginning and end of the region. *E. coli* can synthesize 1700 enzymes. Therefore, this small bacterium has the genes for 1700 different mRNAs.

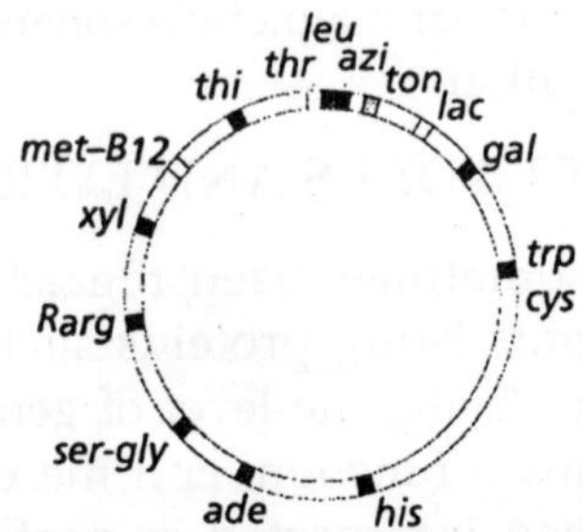

Fig. Partial Gene Map of the Operons, such as Trp and Lac, on a Bacterial Chromosome

Lactose, milk sugar, is split by the enzyme b-galactosidase. This enzyme is inducible, since it occurs in large quantities only when lactose, the substrate on which it operates, is present. Conversely, the enzymes for the amino acid tryptophan are produced continuously in growing cells unless tryptophan is present. If tryptophan is present the production of tryptophan-synthesizing enzymes is repressed.

THE OPERON MODEL

The operon model of prokaryotic gene regulation was proposed by Fancois Jacob and Jacques Monod. Groups of genes coding for related proteins are arranged in units known as operons.

An operon consists of an operator, promoter, regulator, and structural genes. The regulator gene codes for a repressor protein that binds to the operator, obstructing the promoter (thus, transcription) of the structural genes. The regulator does not have to be adjacent to other genes in the operon. If the repressor protein is removed, transcription may occur.

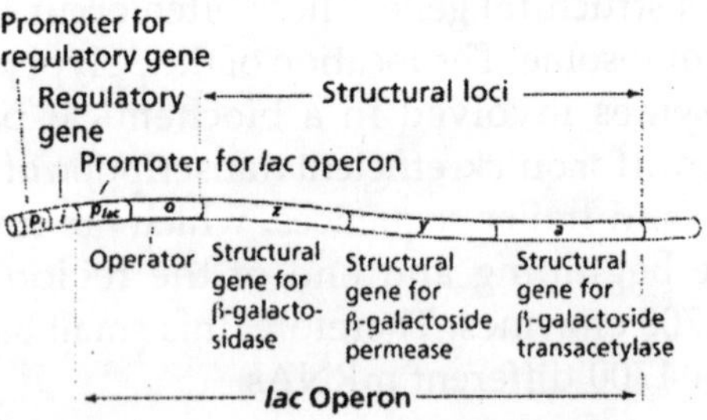

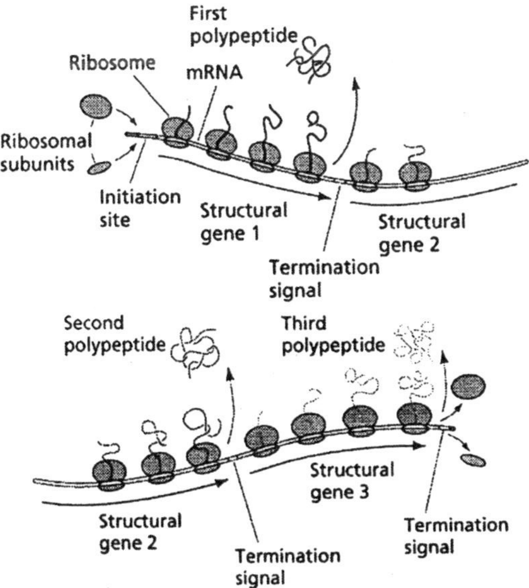

Fig. The Structure and Operation of an Operon

Operons are either inducible or repressible according to the control mechanism. Seventy-five different operons controlling 250 structural genes have been identified for *E. coli*. Both repression and induction are examples of negative control since the repressor proteins turn off transcription.

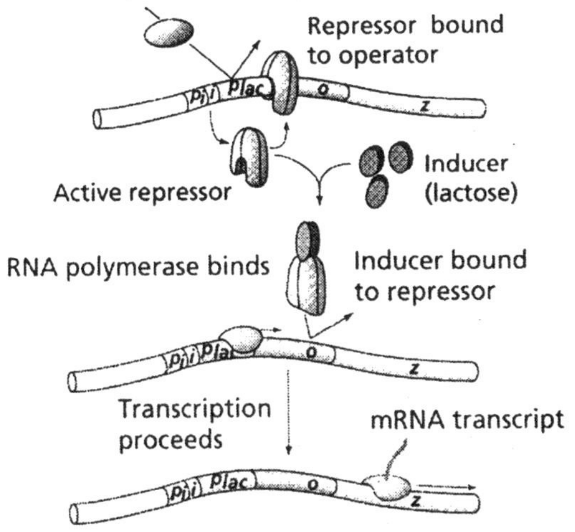

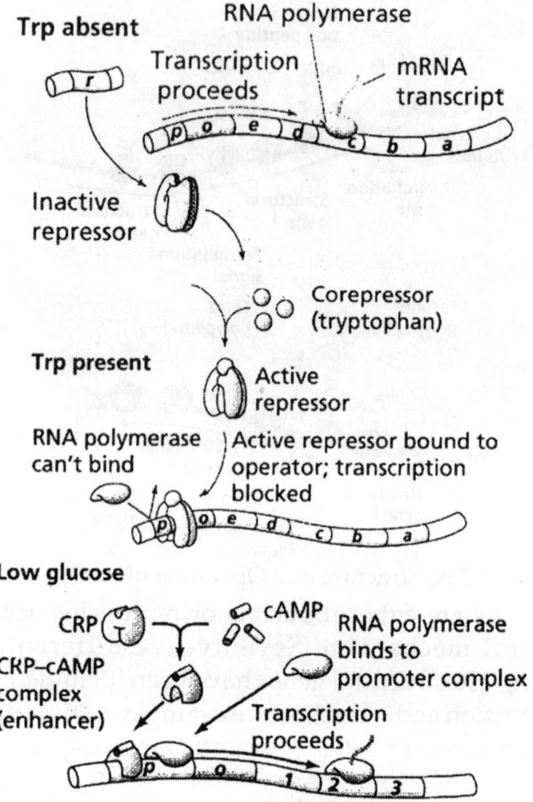

Plasmids, small DNA fragments, are known from almost all bacterial cells. Plasmids carry between 2 and 30 genes. Some seem to have the ability to move in and out of the bacterial chromosome.

An episome is a plasmid incorporated in the bacterial chromosome. Plasmids are self-replicating in a manner like the bacterial chromosome.

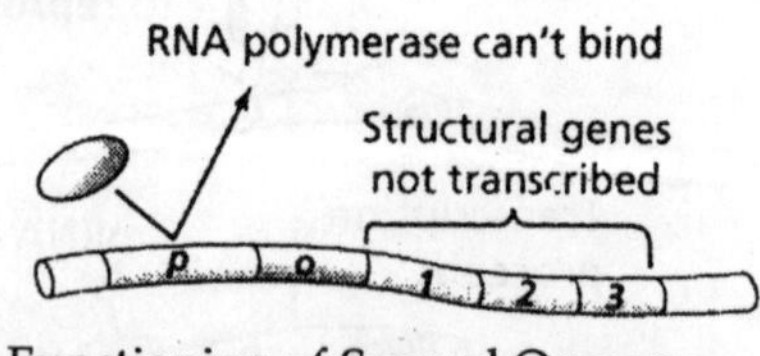

Fig. Functioning of Several Operons

Many plasmids have been recognized for *E. coli*, including the F plasmids ("sex factors") and R plasmids (drug/antibiotic resistance). The F plasmid contains 25 genes, some of which control the production of F pili (proteins which extend from the surface of F^+, or male, cells to the surface of F^-, or female, cells).

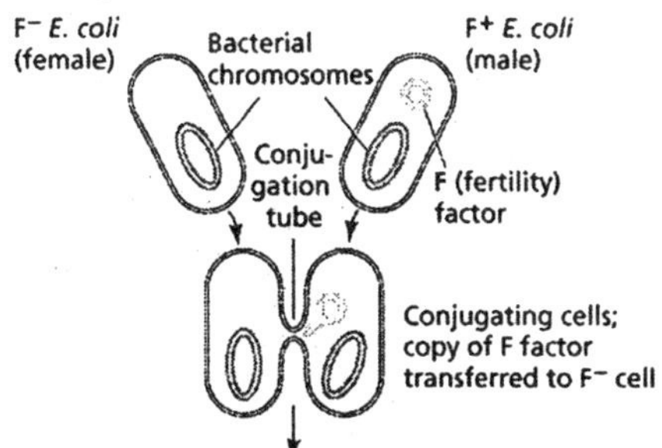

The R plasmid conveys drug resistance on cells having it. As many as 10 resistance genes can be contained on a single R plasmid.

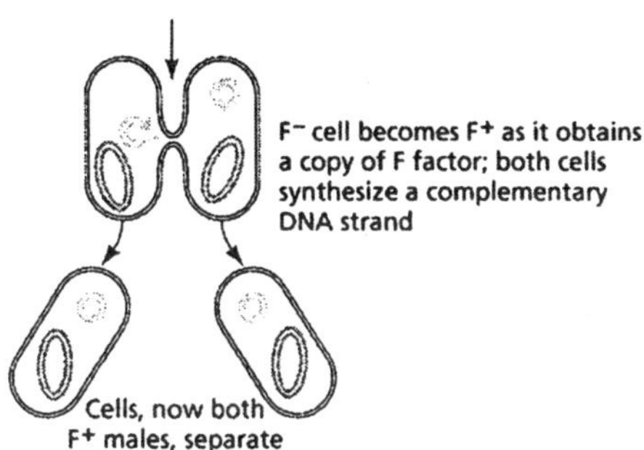

The R plasmids can be transferred to other bacteria of the same species, to viruses, and even to bacteria of different species. Drug (antibiotic) resistance has been found among pathogens causing the diseases typhoid fever, gastroenteritus, plague, undulant fever, meningitis, and gonorrhea.

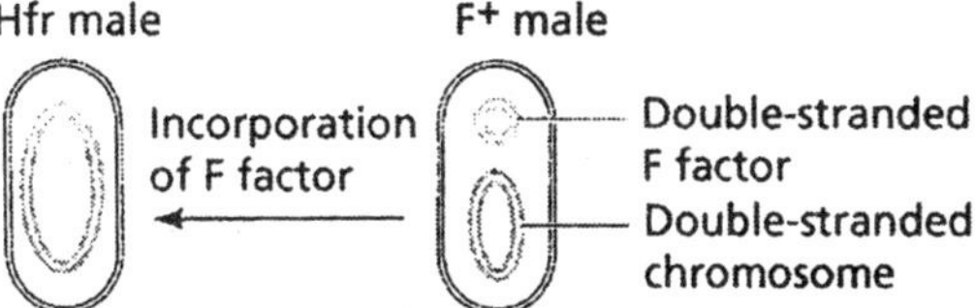

Fig. Conjugation and Exchange of Genetic Material in Bacteria

In addition to the more common modes of transfer, R plasmids may be passed through the cell membrane. The resistance genes appear to operate by either breaking down the antibiotics or by circumventing the block the antibiotic places on a key bacterial metabolic pathway.

VIRUSES

Viruses consist of a nucleic acid (DNA or RNA) enclosed in a protein coat (known as a capsid). The capsid may be a single protein repeated over and over, as in tobacco mosaic virus (TMV). It may also be several different proteins, as in the T-even bacteriophages.

Once inside the cell, the nucleic acid follows one of two paths: lytic or lysogenic.

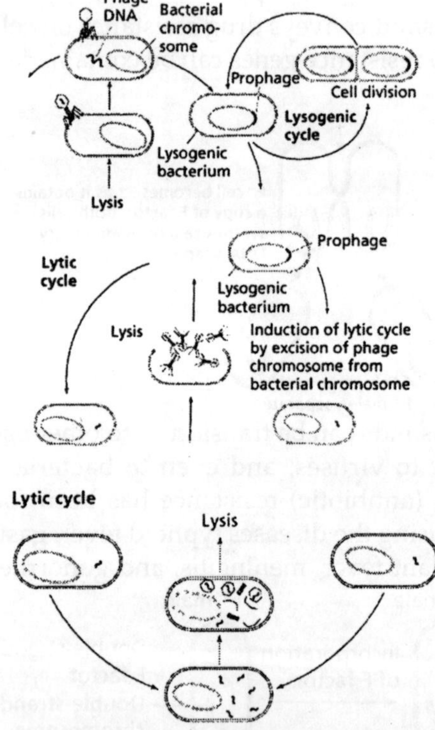

Fig. The Lytic and Lysogenic Phases of a Viral Replication Cycle

Retroviruses, such as Human Immunodifficiency Virus (HIV), also include the enzyme reverse transcriptase with the viral RNA. Reverse transcriptase makes a single-stranded viral DNA copy of the single-stranded viral RNA. The single stranded viral DNA is subsequently turned into a double-stranded DNA.

The lytic cycle occurs when the viral DNA immediately takes over the host cell (remember that viruses are obligate intracellular parasites) and begins making new viruses. Eventually the new viruses cause the rupture (or lysis) of the cell, releasing those new viruses to continue the infection cycle. The lysogenic cycle occurs when the viral DNA is incorporated into the host DNA as a prophage. When the cell replicates the prophage is passed along as if it were host DNA. Sometimes the prophage can emerge from the host chromosome and enter the lytic cycle spontaneously once every 10,000 cell divisions. Ultraviolet light and x-rays may also trigger emergence of the prophage.

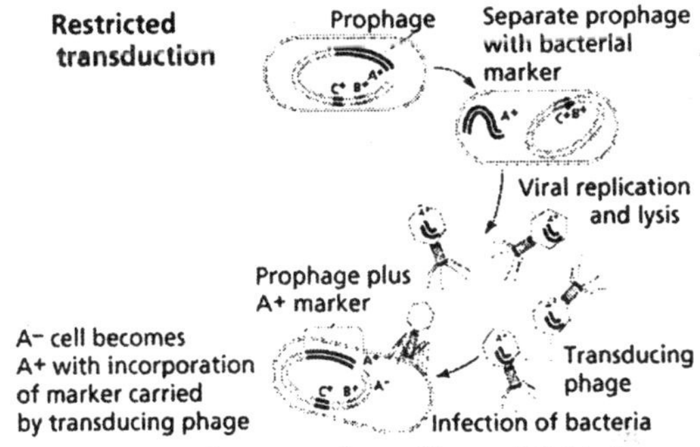

Transduction is the transfer of host DNA from one cell to another by a virus. Some bacteriophages are temperate since they tend to go lysogenic rather than lytic. These types of viruses are able to transduce fragments of the host DNA.

Transposons are DNA fragments incorporated into the chromosomal DNA. Unlike episomes and prophages, transposons contain a gene producing an enzyme that catalyzes insertion of the transposon at a new site. They also have repeated sequences 20-40 nucleotides in length at each end. Insertion sequences are short (600-1500 base pairs long) simple transposons that do not carry genes beyond those essential for insertion of the transposon into *E. coli*. Complex transposons are much larger and carry additional genes.

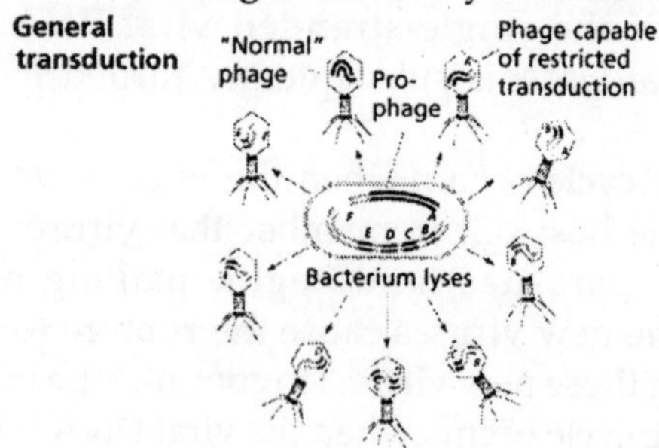

Fig. Induction of Transduction by Viruses in Bacteria

Genes incorporated in a complex transposon are known as jumping genes since they can move about on the chromosome (even from chromosome to chromosome). Often the complex transposons are flanked by simple transposons.

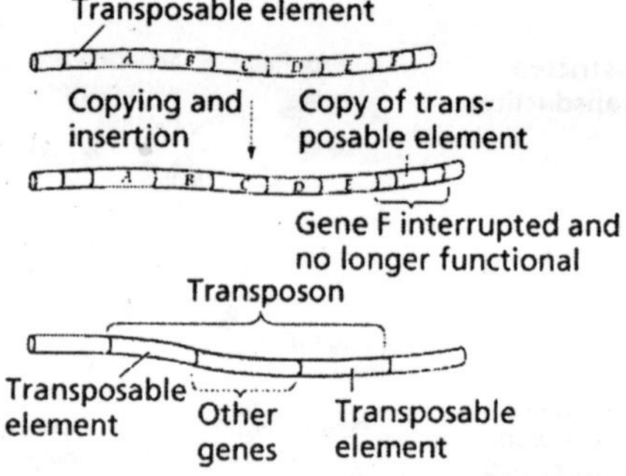

Fig. Transposons and their Relationship to other Genes

THE EUKARYOTIC CHROMOSOME

The eukaryotic chromosome consists of DNA and proteins that appear to play a major role in regulation of eukaryote genes. The DNA of each chromosome is a long single molecule of double stranded DNA. Eukaryotic DNA comes in two forms. Chromatin is the uncoiled form of DNA and is over 50% protein. Chromosomes are coiled DNA/protein that form during the early stages of cell division.

The proteins associated with DNA are collectively known

as histones. They are relatively short polypeptides which are positively charged (basic) and thus are attracted to the negatively charged (acidic) DNA. Histones are synthesized in quantity during the S-phase of the cell cycle.

One function of theses proteins seems to be the folding and packaging of DNA into chromosome form: the 2 m of DNA in a human cell are packaged into 46 chromosomes with a combined length of 200nm (a nm remember is 10-6m). Some 90 million molecules of histones occur in a single cell, with the majority (30 million) being H1 histones. Five types of histone are known (H1, H2A, H2B, H3, and H4); with the exception of H1 most eukaryote histones are very similar.

A nucleosome is the fundamental packing unit of eukaryotic DNA. The core consists of two molecules each of H2A, H2B, H3, and H4; around which the DNA is wound twice. The H1 histone is outside the core. Between 150-200 nucleotide pairs are associated with the core and linker DNA. This level of packing is known as "beads on a string".

The next level are known as the 30nm strand, whose details of organization are not yet well known. 30 nm strands are further condensed into 300nm wide looped domains. Looped domains are part of condensed sections of chromosomes (the chromosome being 1400nm wide at Metaphase I).

REPLICATION OF THE EUKARYOTIC CHROMOSOME

Nucleotide triphosphates (in the forms ATP, GTP, CTP and TTP) are assembled according to the semi-conservative model. Other details of DNA replication are consistent with what we know for prokaryotes.

Eukaryotic DNA has many replication forks and also bidirectional synthesis, contrasting to the unidirectional rolling theta prokaryotic method. Eukaryotic DNA is synthesized much slower than prokaryotic DNA, in humans 50 nucleotides per second per replication fork. After replication the new DNA is immediately associated with histones.

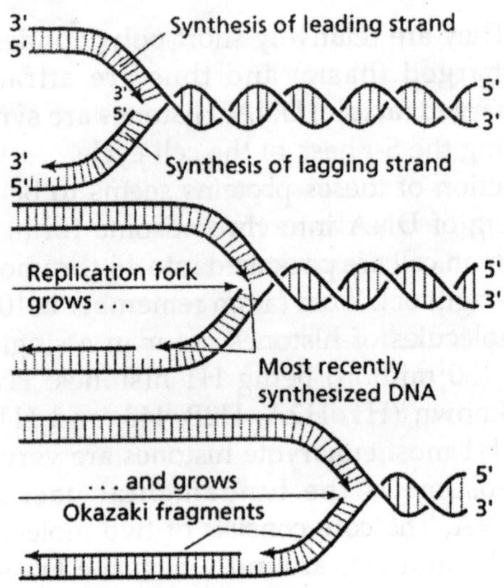

Fig. Replication Fork and its Growth in a Eukaryote.

REGULATION OF EUKARYOTIC GENE EXPRESSION

Eukaryotic gene regulation, especially in multicellular organisms, is complicated by the process of development unique to multicellular organisms. Each multicellular organism begins as a single-celled zygote which divides by mitosis. Cells differentiate into functional types by using some genes but ignoring others.

Homeobox genes establish the body plan and position of organs in response to gradients of regulatory molecules. The timing of certain gene expressions seems to follow a sequence, such as the production of different types of fetal hemoglobins by mammalian red blood cells, which switch to adult hemoglobin sometime after birth. Clearly the inactivation of certain genes occurs in every adult cell; therein lies the cure for cancer, old age, etc.

TYPES OF CHROMATIN

Heterochromatin stains more strongly and is a more condensed chromatin. Euchromatin stains weakly and is more open (less condensed). Euchromatin remains dispersed (uncondensed) during Interphase, when RNA transcription occurs. Some regions of heterochromatin appear to be structural (as in the heterochromatin near the centromere region).

Barr bodies, irreversibly inactivated X-chromosomes, are also condensed heterochromatin. Other heterochromatin regions vary from cell to cell. As the cell differentiates, the proportion of heterochromatin to euchromatin increases, reflecting increased specialization of the cell as it matures.

Loops (or puffs) in insect chromosomes are areas of active RNA synthesis, suggesting again the functional genes are located in open areas of the chromatin (or, the euchromatin). Eukaryotes also have specific binding proteins working in a similar fashion to prokaryotic mechanisms, however the eukaryotes, as one would expect, have a much more complicated process.

THE EUKARYOTIC GENOME

We use the term genome to refer to all of the alleles possessed by an organism (or by a population, species, or larger taxonomic group). While the amount of DNA for a diploid cell is constant within a species, the differences can be great between species.

Humans have 3.5 X 10^9 base pairs, *Drosophila* has 1.5 X 10^8, toads have 3.32 X 10^9, and salamanders have 8 X 10^{10} base pairs per haploid genome. Much of the DNA in each cell either has no function or has a function not yet known. Eukaryotes have only 10% of their DNA coding for proteins. Humans may have a little as 1% coding for proteins. Viruses and prokaryotes use a great deal more of their DNA.

Almost half the DNA in eukaryotic cells is repeated nucleotide sequences. Protein-coding sequences are interrupted by non-coding regions. Non-coding interruptions are known as intervening sequences or introns. Coding sequences that are expressed are exons.

Most, but not all structural eukaryote genes contain introns. Although transcribed, these introns are excised (cut out) before translation (a seemingly energy inefficient process). The number of introns varies with the particular gene, even occurring in tRNAs, rRNAs and viral genes! Generally the more complex and recently evolved the organism, the more numerous and larger the introns.

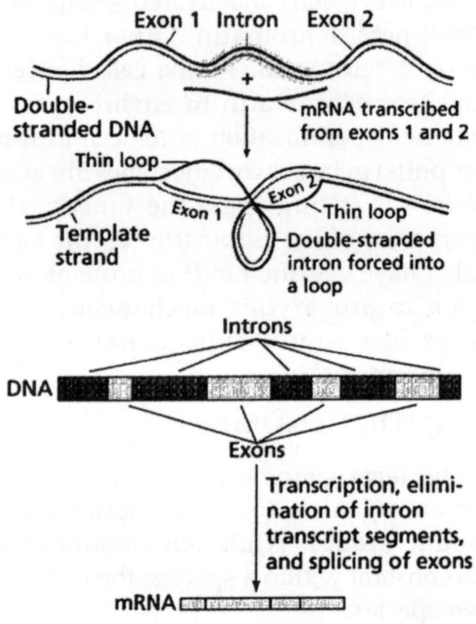

Fig. Introns and Exons

Which came first: continuous genes lacking introns or interrupted genes containing introns? Introns have been hypothesized to promote genetic recombination (via crossing-over), thus speeding up the evolution of new proteins. Exons are also thought to code for different functional regions of proteins. Gene families are made up of similar, but not identical, genes. The globin family is the best studied gene family. Hemoglobin consists, in humans, of 2 a-chains and 2 b-chains clustered about a common heme. Human beta-globin genes are scattered at five loci on human chromosome 11. These genes are expressed sequentially during development, and are similar with same-length introns in similar positions in each gene. Some of the genes are inactivated copies, others are functional only during certain phases of development. The actin family of genes also exhibits a similar pattern.

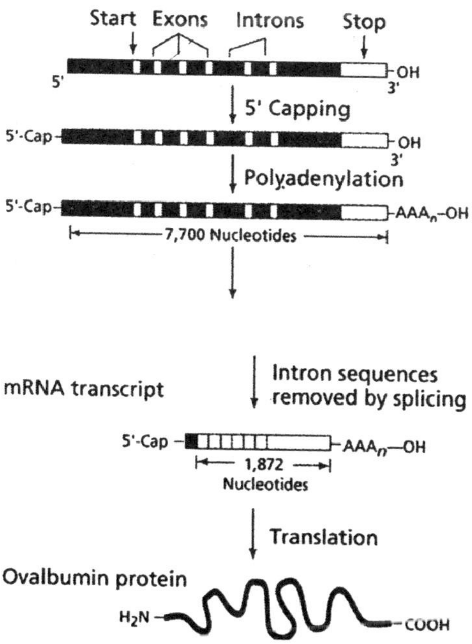

TRANSCRIPTION AND PROCESSING OF mRNA

The process of transcription in eukaryotes is similar to that in prokaryotes, although there are some differences. Eukaryote genes are not grouped in operons as are prokaryote genes. Each eukaryote gene is transcribed separately, with separate transcriptional controls on each gene.

Whereas prokaryotes have one type of RNA polymerase for all types of RNA, eukaryotes have a separate RNA polymerase for each type of RNA. One enzyme for mRNA-coding genes such as structural proteins. One enzyme for large rRNAs. A third enzyme for smaller rRNAs and tRNAs.

Prokaryote translation begins even before transcription has finished, while eukaryotes have the two processes separated in time and location (remember the nuclear envelope). After eukaryotes transcribe an RNA, the RNA transcript is extensively modified before export to the cytoplasm. A cap of 7-methylguanine (a series of an unusual base) is added to the 5' end of the mRNA; this cap is essential for binding the mRNA to the ribosome. A string of adenines (as many as 200 nucleotides known as poly-A) is added to the 3' end of the mRNA after transcription.

The function of a poly-A tail is not known, but it can be used to capture mRNAs for study. Introns are cut out of the message and the exons are spliced together before the mRNA leaves the nucleus. There are several examples of identical messages being processed by different methods, often turning introns into exons and vice-versa. Protein molecules are attached to mRNAs that are exported, forming ribonucleoprotein particles (mRNPs) which may help in transport through the nuclear pores and also in attaching to ribosomes.

ANTIBODY-CODING GENES

Antibodies are globular complex proteins made by multicellular individuals in response to a specific antigen (a foreign substance that has labels saying "I am a stranger"). The cells making antibodies are lymphocytes, better known as white blood cells. Antibodies immobilize and destroy their specific antigens. A mouse can make 10,000,000 antibodies, more than the total genes of the mouse would suggest. Antibody proteins are composed of two long and two short chains. Each species has a constant region characteristic for the species and type of antibody. The other region is the variable in which antigen-specific amino-acids are located.

Susumu Tonegawa tested (and proved) an old hypothesis that the constant and variable regions of antibodies were coded for by different genes. Tonegawa's work further demonstrated that gene fragments in embryos are rearranged to form the variable (functional) genes.

VIRUSES AND EUKARYOTES

The viruses of eukaryotes are similar to prokaryote-infecting viruses. Proviruses are viral DNA integrated into the host cell. Some of the DNA viruses can either initiate an infection (lytic in prokaryotes) cycle or can form proviruses. Simian Virus 40 (SV40) causes cancer in hamsters but not in its normal hosts. SV40 can introduce new functional genes into the host DNA, as can a number of other viruses.

Retroviruses can also insert their nucleic acid into host DNA via the reverse transcriptase mode mentioned previously. Reverse transcriptase is carried with the RNA into the host cell. In the process of making the cDNA strand for the viral RNA, the enzyme also makes long terminal repeats (LTRs) sequences at the terminal ends of the cDNA. These LTRs may also make insertion of the viral DNA into the host DNA easier. The inserted viral DNA makes RNA transcripts which are packaged with viral protein coats and reverse transcriptase.

The viral DNA may, depending on its insertion point, cause mutations of the host DNA. Most viral DNA insertions do not damage the host, but rather become part of the host genome and can be passed on if they have managed to infect a germ-line cell. From 0.5 to 1.0% of mouse DNA may be of viral origin.

EUKARYOTIC TRANSPOSONS

Eukaryotic transposons resemble prokaryotic transposons in many features. Many eukaryotic transposons are first copied to RNA and then back to DNA before being inserted in a new location.

GENES, VIRUSES AND CANCER

Cancer is a disease in which cells escape the restraints on normal cell growth. Cancer is an inheritable disease (at least from cell to daughter cells). Once a cell has become cancerous, all of its descendant cells are cancerous. Gross chromosomal abnormalities are often visible in cancerous cells. Most carcinogens (cancer-generating factors) are also mutagens (mutation-generating factors). Oncogenes are genes resembling normal genes but in which something has gone wrong, resulting in a cancer. Fifty oncogenes have thus far been discovered.

Viruses seem able to cause cancer in three ways. Presence of the viral DNA may disrupt normal host DNA functions. Viral proteins needed for virus replication may also affect normal host gene regulation. Since most cancer-causing viruses are retroviruses, the virus may serve as a vector for oncogene insertion. Transfers of genes between eukaryotic cells will allow doctors, who have historically been limited to phenotypic cures, to attack disease at the genotypic level. SV40 virus has been used to inject the rabbit beta-globin gene into monkeys. Viruses can thus serve as a possible vector to place healthy (non-mutated) alleles into eggs.

GENE REGULATION IN EUKARYOTES

Like prokaryotes, eukaryotic organisms do not want to express all of their genes all of the time. Given the complexity of multicellular eukaryotes, gene regulation in these organisms needs to be very complex.

This module provides a brief overview of the various levels of regulation of gene expression in eukaryotes, and takes a look at the basics of transcriptional regulation; a detailed look at eukaryotic transcriptional regulation is beyond the scope of this course.

THE NEED FOR GENE REGULATION IN EUKARYOTES

Eukaryotes need to regulate their genes for different reasons than prokaryotes. In prokaryotes, gene regulation allowed them to respond to their environment efficiently and economically. While eukaryotes can respond to their environment, the main reason higher eukaryotes need to regulate their genes is cell specialization. Whereas prokaryotes are (relatively speaking) simple, unicellular organisms, multicellular eukaryotes consist of hundreds of different cell types, each differentiated to serve a different specialized function.

Each cell type differentiates by activating a different subset of genes. Because of the multitude of cell types, the regulation of gene expression required to bring about such differentiation is necessarily complex. One way this complexity is demonstrated is in multiple levels of regulation of gene expression.

LEVELS OF REGULATION

Before we discuss the specifics of regulation, it is necessary to understand that "gene expression" covers the entire process from transcription through protein synthesis. The final measure of whether or not a gene is "expressed" is if the protein is produced, because it is protein that will ultimately carry out the function specified by the gene.

We've seen numerous examples of how eukaryotic cells are more complex than prokaryotic cells. One obvious example of this is the presence of a nucleus in eukaryotic cells, which separates transcription from translation in a way not seen in prokaryotes. Furthermore, eukaryotic transcripts must be processed before they can be translated. Here is a diagram outlining the steps involved in the production of a protein in eukaryotic cells:

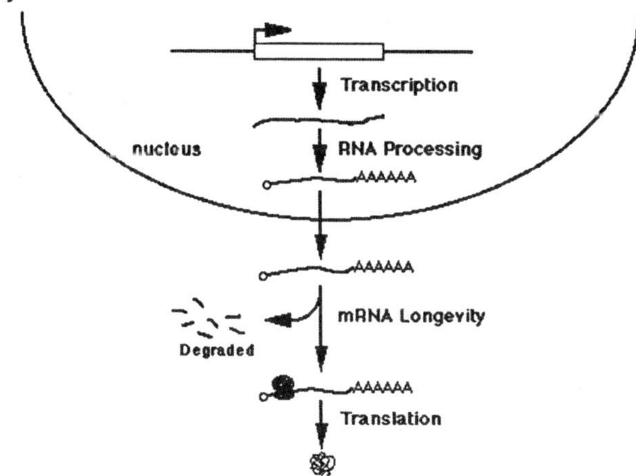

Regulation can occur at any point in this pathway; specifically, it occurs at the levels of transcription, RNA processing, mRNA lifetime (longevity), and translation. Each of these types of regulation will be considered in turn.

REGULATION OF RNA PROCESSING

After transcription, the RNA must be processed before it can be translated. As described elsewhere, RNA processing involves addition of a 5' cap, addition of a 3' poly (A) tail, and removal of introns. This processing represents another level of regulation of gene expression, particularly in regard to splicing out of introns. Regulation can be of two types:

- Whether an RNA gets processed;
- Which exons are retained in the mRNA.

The first type of regulation can determine whether or not an mRNA gets translated. If an RNA is not processed, it will not be transported out of the nucleus, and will not be translated.

The second type of regulation can affect the function of the protein produced. Some genes have exons that can be exchanged in a process known as exon shuffling. For example, a gene with four exons might be spliced differently in two different cell types. In cell 1, exons 1, 2, and 4 would be used in the mRNA:

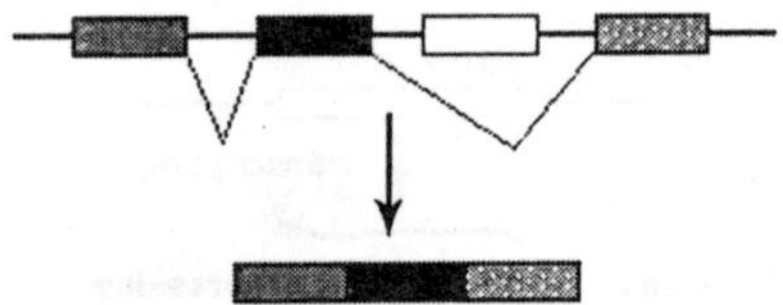

In cell 2 on the other hand, exons 1, 3, and 4 would be used:

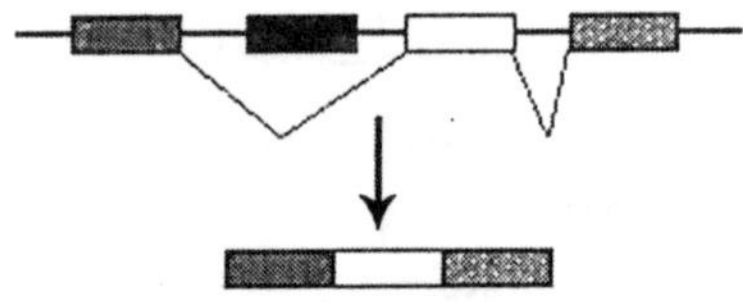

In each of these cases, the polypeptide produced could have a different function. In mammals, for example, the calcitonin gene produces a hormone in one cell type, and a neurotransmitter in another cell type, due to exon shuffling. In *Drosophila*, alternate splicing of the *sex-lethal* RNA can produce an mRNA encoding a functional polypeptide, or one with a premature stop codon that encodes a short, nonfunctional polypeptide.

REGULATION OF RNA LONGEVITY

Imagine two mRNA molecules: one lasts for five minutes

in the cytoplasm before being degraded, while the other one manages to linger for an hour before being degraded. If both are translated continually while they exist, it is obvious that more of the second polypeptide will be produced than the first. This is the principle behind regulation of RNA longevity.

mRNAs from different genes have their approximate lifespan encoded in them; this serves to help regulate how much of each polypeptide is produced. The information for lifespan is found in the 3' UTR. The sequence AUUUA, when found in the 3' UTR, is a signal for early degradation (and therefore short lifetime). The more times the sequence is present, the shorter the lifespan of the mRNA. Because it is encoded in the nucleotide sequence, this is a set property of each different mRNA; the longevity of an mRNA can't be varied.

REGULATION OF TRANSLATION

Whether or not an mRNA molecule is translated can be regulated as well. The various mechanisms of translational regulation are incompletely understood, but there are many documented examples (particularly in embryonic development) of mRNA molecules that are present routinely, but are only translated under certain circumstances. For example, many animals sequester large amounts of mRNA in their eggs, and those mRNA molecules are not translated unless the egg is fertilized.

REGULATION OF TRANSCRIPTION

Whether or not a gene is transcribed is the major way that gene expression is regulated in eukaryotes, as it was in prokaryotes. There are some major differences between transcriptional regulation in prokaryotes and eukaryotes. For one thing, because of the complexity of eukaryotic patterns of gene expression, each eukaryotic gene needs its own promoter. In other words, eukaryotic genes are not organized into operons. Another difference is that prokaryotic genes are regulated primarily by repressors. Although repressors occasionally play a role in eukaryotes, eukaryotic genes are primarily regulated by transcriptional activators. These activators are transcription factors.

REGULATORY ELEMENTS OF EUKARYOTIC GENES

As discussed in the module on transcription, eukaryotic genes have promoters that are recognized by basal transcription factors (such as TFIID).

In addition to the promoter, eukaryotic genes have one or more enhancers. These are DNA sequences associated with the gene being regulated, and whereas the promoter is responsible for initiating low levels of transcription and determining the transcription start site, enhancers are responsible for increasing ("enhancing") transcription levels, and they are responsible for regulating cell- or tissue-specific transcription (i.e. the transcription responsible for differentiation). There are some other basic differences between enhancers and promoters, which are outlined in the module on transcription.

Enhancers function by being recognized and bound to by transcription factors. These are not the basal-type transcription factors (such as TFIID) discussed elsewhere. These are specialized transcription factors, of which there are very many types (all are proteins). A very large number of enhancer elements has been identified and characterized, and each different enhancer has its own transcription factor that it binds to.

TISSUE-SPECIFIC GENE EXPRESSION

How is a gene turned on in one cell type, and not in another? It depends primarily on whether the transcription factor for the gene's enhancer is active or not in a cell. To illustrate this, let's consider two different genes - one is regulated by enhancer A, which is recognized by transcription factor A, and the other gene is regulated by enhancer B, which is recognized by transcription factor B. In one cell type (let's say muscle, for the sake of argument), transcription factor A might be active whereas transcription factor B might not. In such a cell, only the first gene would be transcribed:

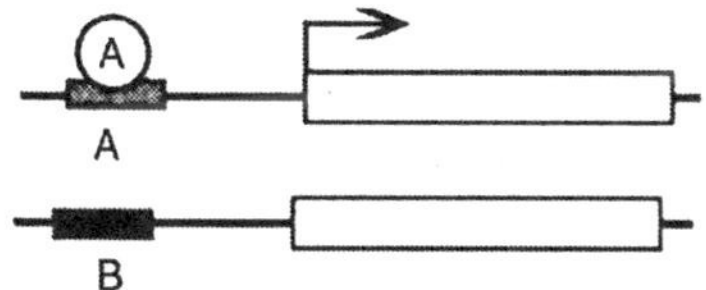

In another cell type (e.g. epidermis), transcription factor B would be active, and transcription factor A would not. In this cell, only the second gene would be transcribed:

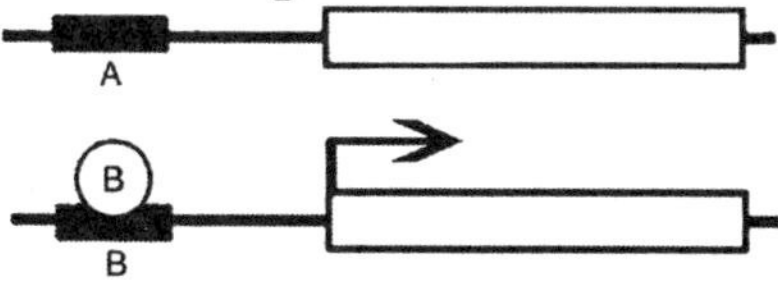

So how is it determined which transcription factors are active in which cell types? The answer to this question is very complicated, and not completely understood. However, there are a number of mechanisms by which transcription factors can be regulated:

Often, the presence or absence of a transcription factor in a cell is the determining step. If the factor is present, the gene is transcribed. If not, then the gene is not transcribed. The presence of a transcription factor, of course, depends upon the activity of the gene encoding that transcription factor, which forces us to ask how the gene encoding the transcription factor is regulated, which pretty much brings us back to where we began.

Transcription factors can be activated by environmental signals. For example, virtually all organisms have a set of genes called heat shock genes that encode proteins that help the organism survive heat stress. These genes are activated under conditions of heat stress, under the control of a transcription factor called heat shock transcription factor. This factor is always present, but is only activated when greatly increased temperatures are detected.

Transcription factors can be activated by signals from other cells in the same organism. Such signals include hormones and growth factors. Hormones must bind to a specific receptor on the target cell, and the receptor mediates the cellular effects of the signal. There are two basic mechanisms used, one for steroid hormones, and one for peptide hormones:

Steroid hormones are lipid (actually cholesterol) derivatives, such as testosterone and progesterone. These hormones can cross the cytoplasmic membrane into a cell, where they bind to their specific receptor. Steroid receptors are transcription factors, and when they bind to their ligand, they become activated and initiate transcription of a specific set of genes.

Peptide hormones cannot cross the cytoplasmic membrane, and so must bind to a receptor on the cell surface. When bound to its ligand, these receptors initiate a series of biochemical reactions inside the cell, with the ultimate result being the activation of a transcription factor (often by phosphorylation of the transcription factor), which initiates transcription of a specific set of genes.

DEVELOPMENTAL GENETICS

The primary function of the majority of genes in eukaryotic organisms is to coordinate the development of embryos. This module takes a look at some of the basic principles underlying the role of genes in embryonic development.

In most of the other modules of this course, the effects of genes are examined with regard to their effect on the phenotype of the juvenile or adult organism. In reality, however, the vast majority of these phenotypes are established during embryonic development, because most genes function to generate the pattern (the 'shape') of the developing embryo. Loss of function of a particular gene results in an abnormality in development, which manifests itself as a phenotype in the adult.

In this sense, virtually all of eukaryotic genetics is developmental genetics, even though we haven't considered it as such. The field of developmental biology is concerned with the function of genes in embryogenesis. The central question of developmental biology is the following:

How does a single cell, the fertilized egg (or zygote),

manage to produce an extremely complex adult organism, which is composed not only of trillions of cells, but of thousands of different types of cells (such as nerve cells, muscle cells, etc.)?

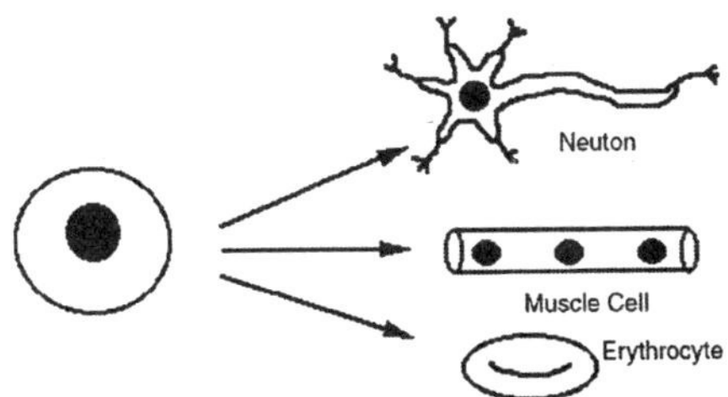

There is a simple answer to this question, and that is that cells differentiate. In other words, different cells become specialized to carry out different functions by following different developmental pathways. But how does this occur? How do cells know what pathway to follow?

THE MOSAIC THEORY OF DEVELOPMENT

One early theory (in the late 1890's) that attempted to explain the process of differentiation was the mosaic theory, as outlined by Wilhelm Roux and August Weismann. This theory proposed that there were determinants that specified the various differentiation pathways. The theory stated that these determinants would all be present in the zygote, but as cell division occurred after fertilization, the determinants would be unequally inherited by the offspring cells.

This is known as qualitative cell division. The set of determinants would thus be divided up until each cell contained only one type of determinant, and that determinant would determine the fate of that cell. When Mendel's work was rediscovered in the early 1900's and the concept of the 'gene' was developed, it was believed that the genes were the determinants that were divided up during cell division.

Evidence was accumulated over sixty years that disproved the notion that genes are divided up in development. The final nail in the coffin of this theory came from nuclear transplantation studies done using amphibians in the early 1960's. In these studies, nuclei were isolated from tadpole intestinal cells (differentiated cells) and injected into eggs that had their own haploid nuclei removed. A small percentage of the injected eggs developed into completely normal adult frogs! These frogs had developed using only the genetic information found in a tadpole intestinal nucleus.

Therefore, a tadpole intestinal nucleus must contain all of the genetic information necessary to allow the differentiation of every cell type in an adult frog. If the mosaic theory were correct, this would not be the case; a tadpole intestinal nucleus would have only the genetic information necessary to cause the differentiation of an intestinal cell.

Note: The frogs produced by this procedure would be genetically identical to the frog from which the intestinal cell nucleus was obtained. In other words, they would be clones. In fact, these experiments comprised the earliest successful cloning of vertebrate organisms. The recent cloning of Dolly the sheep was done for the same reasons as outlined above: demonstrating that differentiated mammalian nuclei (from mammary gland in this case) have all of the genetic information necessary to drive the development of a normal adult organism.

THE THEORY OF DIFFERENTIAL GENE EXPRESSION

If all nuclei in an organism contain the same genetic information (as we know they do), then how does differentiation occur? The explanation for this is found in the Theory of Differential Gene Expression. This theory states that differentiation occurs as a result of expression in a particular cell of only a subset of the total genes present. For example, if a cell expresses only the set of genes that causes muscle differentiation, then that cell will differentiate into a muscle cell.

Alternatively, if the cell expresses only the set of genes that causes spleen cell differentiation, then the cell will differentiate into a spleen cell. This is a fairly simple concept, but it hasn't really answered the question. The question has now become: how do cells activate only a certain set of genes (inactivating all other genes), and how do they know which set of genes to activate? The answer to this question is fairly straightforward as well: the set of genes activated in a cell is dependent on the set of transcription factors found in the cell.

Using the muscle cell example from above, the cell activates the muscle-specific genes because it contains transcription factors that specifically activate the muscle-specific genes. If you think about this for a minute, you'll realise that we still haven't answered the question; we've only changed it again. Now the question becomes: how did the cell come to contain only that specific set of transcription factors? Well, those transcription factors are encoded by genes, and those genes are regulated by other specific transcription factors. Those transcription factors are in turn encoded by other genes, which are regulated by still other transcription factors, etc.

As we can see, there is a hierarchy of genes within each cell. Genes are expressed that encode transcription factors, which activate other genes that encode transcription factors, which activate other genes that bring about differentiation along a specific pathway. How many levels are there to the hierarchy? From what we've seen so far, the hierarchy seems to go on forever. Each set of transcription factors is encoded by genes that are activated by yet another set of transcription factors. It must end somewhere, but where?

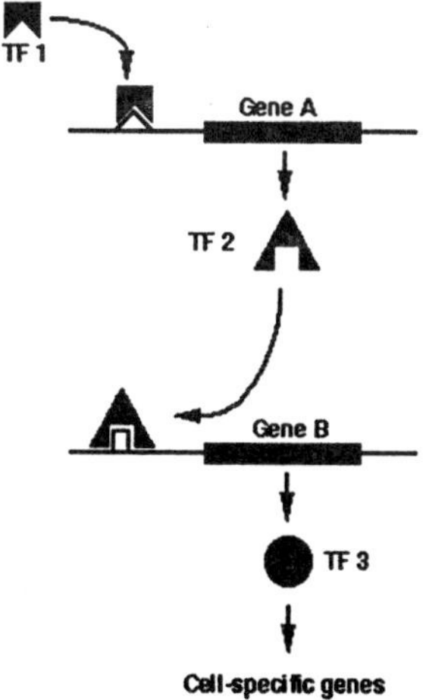

MASTER CONTROL GENES

In one sense, it ends with master control genes. A master control gene is the first gene activated in a hierarchy that leads to differentiation along a particular pathway. Master control genes encode the first transcription factor in a hierarchy; a master control gene product activates the next set of genes that encodes the next set of transcription factors, and the cascade of gene expression has been set in motion.

Let's look again at the muscle cell example considered earlier. The master control 'gene' for muscle development is actually a family of genes called the MyoD gene family. These genes encode HLH-type transcription factors. How do we know these are the master control genes for muscle? Master control genes have a particular property. Because they initiate muscle differentiation, if the MyoD genes are activated in other cell types, they cause those cells to transdifferentiate into muscle. For example, hepatocytes (liver cells) or adipocytes (fat cells) that are caused to express the MyoD genes (using tricks of molecular biology) will change from their normal phenotype into muscle cells.

So how are the master control genes regulated? There are two basic ways that this can happen. One way is through the process of induction. Induction occurs when once cell sends a signal to another cell, telling it to differentiate a certain way. The signal, usually a diffusable protein, causes the recipient cell to activate the appropriate master control gene product. This is how muscle differentiation occurs. Signals from other cells cause the MyoD genes to become active in the cells receiving the signal, and muscle differentiation is initiated.

Chapter 8

Functions of Ribosomes

Ribosomes are cytoplasmic granules composed of RNA and protein, at which protein synthesis takes place. They were first observed by Palade in the electron microscope as dense particles or granules. Upon isolation, they were shown to contain approximately equal amounts of RNA and protein Label. To function actively in protein synthesis, they must be bound into complete ribosomes. We know that ribosomes are but one of the required components necessary for the synthesis of protein.

The others are messenger RNA, which carries the genetic message; soluble RNA, which carries amino acids to be synthesized; and guanosine triphosphate, which is the source of energy. A number of ribosomes may be attached to the same messenger, each manufacturing its own chain of polypeptides, called a polysome. Ribosomes are also found in the mitochondria and chloroplasts of eukaryotic cells. They are always smaller than the 80S cytoplasmic ribosomes, and are comparable to prokaryotic ribosomes in both size and sensitivity to antibiotics, although the sedimentation values vary somewhat in different phyla.

Prokaryotic and eukaryotic ribosomes do not differ in any fundamental way; both preform the same functions by the same set of chemical reactions. The genetic code is the same in all living organism, and it has been demonstrated that eukaryotic ribosomes are able to translate bacterial mRNAs correctly. Eukaryotic ribosomes are much larger that prokaryotic ones and most of their proteins are different. Antibiotics such as chloramphenicol inhibit bacterial but not eukaryotic ribosomes. Protein synthesis by eukaryotic ribosomes in inhibited by cycloheximide. Mitochondrial and chloroplast ribosomes resemble those in bacteria.

They are inhibited by chloramphenicol, and hybrid ribosomes containing one bacterial and one chloroplast ribosome subunit, for example, are fully active in protein synthesis. Hybrid eukaryotic ribosomes containing subunits from both plants and mammals are also active in protein synthesis, but are inactive if one of the subunits is derived from bacteria. Some structural resemblance must exist, however, since reconstruction experiments have shown that two proteins from the large subunit of E. coli can replace the homologous proteins in mammalian ribosomes.

In summary, there is little structural but considerably functional homology between prokaryotic and eukaryotic ribosomes. Cells devote considerable effort to the production of these essential organells. For example, an E. coli cell contains approximately 15,000 ribosomes, each one with a molecular weight of about three million daltons. Ribosomes therefore represent twenty-five percent of the total mass of these bacterial cellsLabel.

STRUCTURE

Ribosomes are tiny particles, about 200 A. It is composed of both proteins and RNA; in fact it has approximately 37 - 62% RNA, and rest are made up of proteins Label. The RNA present in ribosomes are obviously called ribosomal RNA, and they are produced in the nucleolus, which is a prominent globular structure in the nucleus. Thus, the proteins are gene products of themselves, and one ribosome is made up of dozens of genes.

The ribosomes fall into two categories: Those that are free to roam in the cytoplasm, and those that are bound to gigantic, cobwebby organelles made up of membranes, called the endoplasmic reticulum; thus, causing a rough surface. Although, the two kinds of ribosomes play similar roles in translating mRNA to produce proteins, they are very distinct in where its product is located. The ribosomes in the cytoplasm allows its protein to roam about freely, while the bound ribosomes transfer their functional protein into the endoplasmic reticulum. In addition, ribosomes are also located within the mitochondria, and the chloroplast, but are only few in content. This spherical particle of 23nm, is composed of two subunits; a large and small Label.

In Eukaryotes, the co-efficient of ribosomes are 80s, of which is divided into 60s for the large, and 40s for the small subunit. The 60s contain 28s rRNA, with a small fragment that is attached noncovalently and can be released upon heating; a 5.8s, and a very small - 120 nucleated of 5sRNA. Whereas, the 40s subunit has only a single 18s rRNA Label. In prokaryotes, however, the large and small subunits are split into 50s and 30s, making a total of 70s respectively. The 50s has two types of rRNA - a 23s and a 5s. It also has 32 different proteins. On the other hand, the 30s contains a single 16s rRNA, 21 different types of proteins.

To help better understand what the s stands for in rRNA, let us use the prokaryotes as an example. The 50s and 30s refers to the sedimentation coefficient of the two subunits. This coefficient is a measure of the speed with which the particles sediment through a solution when spun in an ultra centrifuge. Thus, the particles with larger coefficient would centrifuge and settle much faster since it is has more mass than the particle with the smaller coefficient. 50s + 30s $\rightarrow$ 70s Note that the two subunits above make up the entire ribosomal molecule which is 70s. The reason the coefficients do not add up is because they are not proportional to the particle weight.

During protein synthesis, ribosomes line up along the mRNA and form a polysome, also called the polyribosome. The mRNA is aligned in the gap between the 2 ribosomal subunits. It is possible that the nascent peptide chain grows through a channel or groove in the large ribosomal subunit. This is predicted to be the case since ribosomes protect a segment of 30-40 amino acids from degradation. Speaking of amino acids, up to 30 ribosomes can attach on one strand of mRNA to form amino acid chains thus leading to protein formation. Ribosomes act as the backbone for many molecules during translation. It provides room for many structures to situate itself thus enhancing protein synthesis.

For example, mRNA inserts itself between the two subunits; the peptidyl transferase complex - the enzyme that allows for the tRNA to break apart from the amino acid on P-site; this enzyme lays across the molecule, between the subunits. It contains the P and the A-site for tRNA binding. Last but not least, the ribosome molecule allows the growing polypeptide chain, to emerge from the back of the structure, thus it is situated perpendicular to the mRNA chain. Ribosomes have a tertiary structure. Ribosomes make up a large part of cells in many species, which leads to protein manufacturing. For example, in E.Coli (bacteria), they make up about 1/4 of the total cell mass.

They are intensely basiphilic (having high affinity for bases). Due to its complex structures, with many proteins and different kinds of RNA, researchers have found it very difficult to study the macro molecular structure of ribosomes, especially for the fact it is quite impossible to observe its crystal using an x-ray diffraction. Thus, scientists have been forced to use other means of study to map the proteins and RNA components in ribosome. Some of these are the cross-linking, immunoelectron microscopy, and low-angle neutron scattering methods. The cross-linking shows the protein arrangement and the types of bonds it forms within itself.

The neutron scattering experiments forms horizontal lines that show the entire structure of ribosome, with its two subunits, and shows where the proteins are arranged in the molecule. The empty regions around the proteins is where the rRNA is located. The immunoelectron microscopy, shows the proposed location of the 16s rRNA molecule of the small subunit, in prokaryotes.

FUNCTION

The ribosomes plays a very important role in protein synthesis, which is the process by which proteins are made from individual amino acids. Without the ribosomes the message would not be read, thus proteins could not be produced. Therefore, ribosomes play a very important role in role in protein synthesis. The primary agent in the process of translating the mRNA into a specific amino acid chain is the ribosome, which consists of two subunits. These subunits are made up of a third and extremely abundant type of RNA, ribosomal RNA (rRNA), and together contain up to eighty-two specific proteins assembled in a precise sequence Label.

The ribosomes constituents must be put together in an extremely precise position and sequence. This assembled ribosome displays a series of small groves, tunnels, and platforms, where the action of protein synthesis occurs Label. There are the active sites, each dedicated to one of the tasks required for translation of mRNA into protein. Proteins being synthesized for export out of the cell, are made by ribosomes attached to the rough endoplasmic reticulum. In contrast, proteins for use by the cell are generally made in the cytoplasm by free ribosomes.

Several of these free ribosomes may attach to a single mRNA molecule, giving rise to the polyribosome or polysome Label. Protein synthesis takes place on polyribosomes (or polysomes) where 80S ribosomes associate with an mRNA coding for a given protein. The number of ribosomes associated in the polysomal chains depends on the size of the mRNA. This is also associated with the size of the protein that is being synthesized. Outside the polyribosome, the ribosomes are dissociated and form a pool of free subunits. Transfer RNAs are also bound to the ribosome.

There are quite a few factors involved in the formation of the initiation complex. These include: GTP, methionine tRNA, an initiation codon in mRNA, 80S ribosomes, and three protein factors Label. The process of protein synthesis begins with the capture of the tRNA, which is carrying an amino acid, by an initiation factor. This binds to a small ribosomal subunit, which occupies one of the active sites in the ribosomes, the P (protein) site. This initiation complex recognized and binds to the 5' end of an mRNA molecule and slides down to the initiation codon, which is always an AUG sequence of amino acids.

The large subunit of the ribosome now joins the complex.

A second tRNA is now brought into the ribosome by the elongation factor. If the anticodon of the tRNA pairs with the next codon of the message, the tRNA occupies the A (acceptor) site on the ribosome.

This positions the second amino acid adjacent to the initiation methionine. Then an enzyme, peptidyl transferase, which is part of the large ribosomal subunit mediates the separation of the first amino acid from its tRNA and the formation of a peptide bond between the initial methionine and the amino acid is formed.

The P site is now occupied by an uncharged tRNA molecule label. The ribosome will now move down the mRNA by one codon, a process known as translocation. This movement shifts the growing polypeptide chain to the P position, and results in an empty A site, where a new charged tRNA can enter and pair, by forming a hydrogen bond between the codon and the anticodon. This holds the tRNA into place long enough for an even more stable binding to occur Label. The uncharged tRNA that previously occupied the P site is booted out of the ribosome and will be recharged and recycled by the cell.

The energy needed for this process is supplied by the hydrolysis of guanosine triphosphate (GTP). The process then continues along the length of the mRNA, until the first stop codon is encountered. At that point the action of a termination factor releases the completed protein from the last tRNA and the ribosome dissociates into its component parts. Another function of the ribosomes occurs in the relation to the neuron and axons. The cell body of a typical large neuron contains vast numbers of ribosomes. Although dendrites often contain some ribosomes, there are no ribosomes in the axon, and its protein must therefore be provided by the many ribosomes in the cell body.

FUNCTION LINKAGE

RNA polymerase is the enzyme that directs transcription, which is the process by which the mRNA copy of a gene is synthesized. Transcription follows the same rules of base pairing as DNA replication. This base pattern ensure that an RNA transcript is a faithful copy of the gene. There are three stages of transcription: initiation, elongation, and termination. During initiation, the enzyme recognizes a promoter region, which lies upstream from the gene. The polymerase binds tightly to the promoter and causes localized melting, or separation of the two DNA strands within the promoter. Then the polymerase starts building the RNA chain.

Ribonucleoside triphosphates such as ATP, GTP, CTP and UTP are the building blocks the polymerase uses for this job Label. After the first nucleotide is in place, the polymerase binds the second nucleotide, joining it to the first. This forms the initial phosphodiester bond in the RNA chain. The second stage is elongation, where the RNA polymerase directs the sequential binding of ribonucleotides to the RNA chain. As it does this, it moves along the DNA template and the melted DNA moves with it.

This melted region exposes the bases of the template DNA one-by-one so that they can pair with the bases of the incoming ribonucleotides. As soon as the transcription machinery passes, the two DNA strands wind around each other again, reforming the double helix. Only enough separation will occur so that the polymerase can read the DNA template Label. The final stage is termination, which allows the termination of transcription. These work in conjunction with RNA polymerase, and is sometimes aided by another protein, to loosen the association between RNA product and DNA template.

So the RNA dissociates from the RNA polymerase and DNA, thus terminating transcription. Transcription is very important in that it is the only step in expression of the genes for rRNAs. It is also important to ribosomes, because it sets up the RNA in a 5' to 3' sequence, which allows the ribosomes to read the message 5' to 3'. Without transcription of a gene, the ribosome would not be able to translate an mRNA, thus not allowing it to be separated into its component parts. If this does not occur, then translation will also not occur, thereby affecting the function and role of ribosomes in translation.

REGULATION LINKAGE

Ribosomes are used by all living cells to synthesize proteins. This synthesis can be inhibited by antibiotics, which can have an effect on the organism. Some antibiotics target specific subunits of the ribosome or may target the entire ribosome completely. Streptomycin, for example, can target the 70s ribosome in some prokaryotes and cause adverse effects on the cell of the host. An antibiotic that affects the 30s ribosome is tetracycline.

It can prevent the addition of amino acids to the growing polypeptide chain by interfering with the attachment of the tRNA onto the ribosome. The 50s ribosome may also be targeted by erythromycin, which blocks the translocation reaction on ribosomes. Other antibiotics that interfere with the ribosome to synthesize proteins include chloramphenicol, rifamycin, puromycin, cycloheximide, and anisomycin.

STRUCTURE LINKAGE

Antibiotics are drugs produced by bacteria and fungi. These molecules function as drugs used in the chemotherapy of infectious disease, and follow the principle of selective toxicity. Selective toxicity follows the principle of using drugs that kill the harmful microorganism without damaging the host. As a result of its toxicity, antibiotics can affect the ribosomal structure, inhibiting protein synthesis. For instance, let us take the 70s ribosome of prokaryotes; antibiotics can target this structure and can adverse the effects on the cells of the host. Among the antibiotics that interfere with protein synthesis are chloramphenicol, erythromyocin, streptomycin, and the tetracycline.

This paper will be focused on these four antibioticsand its role played in the effect of the ribosome structure, thus leading to the change in protein synthesis. For instance, the chloramphenicol reacts with the 50s structure of the 70s prokaryote ribosome, by inhibiting the formation of the peptide bonds in the growing polypetide chain. Erythromyocin, the second antibiotic, also reacts with the same structure as chloromphenicol. However, it has a very narrow range of activity, since it affects mostly the gram-positive bacteria.

The other two antibiotics attract the 30s structure of the 70s prokaryotic ribosome. The tetracycline interferes with the attachment of the tRNA, which carries the amino acids, to the ribosome, thus preventing the addition of amino acids to the growing polypeptide chain. One unique aspect of tetracycline is that it cannot penetrate well into the mammalian cells, therefore, it does not interfere with the mammalian ribosomes. Aminoglycoside antibiotics, a type of strept-omycin, changes the shape of the 30s structure of the 70s prokaryotic ribosome, thus interfering with the initial stage of protein synthesis. This in turn, causes the misreading of the genetic code on the mRNA.

MESSENGER RNA

Messenger ribonucleic acid (mRNA) is a molecule of RNA encoding a chemical "blueprint" for a protein product. mRNA is transcribed from a DNA template, and carries coding information to the sites of protein synthesis: the ribosomes. Here, the nucleic acid polymer is translated into a polymer of amino acids: a protein. In mRNA as in DNA, genetic information is encoded in the sequence of four nucleotides arranged into codons of three bases each.

Each codon encodes for a specific amino acid, except the stop codons that terminate protein synthesis. This process requires two other types of RNA: transfer RNA (tRNA) mediates recognition of the codon and provides the corresponding amino acid, while ribosomal RNA (rRNA) is the central component of the ribosome's protein manufacturing machinery.

mRNA

The brief existence of an mRNA molecule begins with transcription and ultimately ends in degradation. During its life, an mRNA molecule may also be processed, edited, and transported prior to translation. Eukaryotic mRNA molecules often require extensive processing and transport, while prokaryotic molecules do not.

Transcription

During transcription, RNA polymerase makes a copy of a gene from the DNA to mRNA as needed. This process is similar in eukaryotes and prokaryotes. One notable difference, however, is that eukaryotic RNA polymerase associates with mRNA processing enzymes during transcription so that processing can proceed quickly after the start of transcription. The short-lived, unprocessed or partially processed, product is termed *pre-mRNA*; once completely processed, it is termed *mature mRNA*.

Eukaryotic pre-mRNA Processing

Processing of mRNA differs greatly among eukaryotes, bacteria and archea. Non-eukaryotic mRNA is essentially mature upon transcription and requires no processing, except in rare cases. Eukaryotic pre-mRNA, however, requires extensive processing.

5′ Cap Addition

A *5′ cap* (also termed an RNA cap, an RNA 7-methylguanosine cap or an RNA m^7G cap) is a modified guanine nucleotide that has been added to the "front" or 5' end of a eukaryotic messenger RNA shortly after the start of transcription. The 5' cap consists of a terminal 7-methylguanosine residue which is linked through a 5'-5'-triphosphate bond to the first transcribed nucleotide. Its presence is critical for recognition by the ribosome and protection from RNases. Cap addition is coupled to transcription, and occurs co-transcriptionally, such that each influences the other. Shortly after the start of transcription, the 5' end of the mRNA being synthesized is bound by a cap-synthesizing complex associated with RNA polymerase. This enzymatic complex catalyzes the chemical reactions that are required for mRNA capping. Synthesis proceeds as a multi-step biochemical reaction.

Splicing

Splicing is the process by which pre-mRNA is modified

to remove certain stretches of non-coding sequences called introns; the stretches that remain include protein-coding sequences and are called exons. Sometimes pre-mRNA messages may be spliced in several different ways, allowing a single gene to encode multiple proteins. This process is called alternative splicing. Splicing is usually performed by an RNA-protein complex called the spliceosome, but some RNA molecules are also capable of catalyzing their own splicing.

Editing

In some instances, an mRNA will be edited, changing the nucleotide composition of that mRNA. An example in humans is the apolipoprotein B mRNA, which is edited in some tissues, but not others. The editing creates an early stop codon, which upon translation, produces a shorter protein.

POLYADENYLATION

Polyadenylation is the covalent linkage of a polyadenylyl moiety to a messenger RNA molecule. In eukaryotic organisms, most messenger RNA (mRNA) molecules are polyadenylated at the 3' end. The poly(A) tail and the protein bound to it aid in protecting mRNA from degradation by exonucleases. Polyadenylation is also important for transcription termination, export of the mRNA from the nucleus, and translation. mRNA can also be polyadenylated in prokaryotic organisms, where poly(A) tails act to facilitate, rather than impede, exonucleolytic degradation.

Polyadenylation occurs during and immediately after transcription of DNA into RNA. After transcription has been terminated, the mRNA chain is cleaved through the action of an endonuclease complex associated with RNA polymerase. The cleavage site is characterized by the presence of the base sequence AAUAAA near the cleavage site. After the mRNA has been cleaved, 80 to 250 adenosine residues are added to the free 3' end at the cleavage site. This reaction is catalyzed by polyadenylate polymerase. Just as in alternative splicing, there can be more than one polyadenylation variant of a mRNA.

TRANSPORT

Another difference between eukaryotes and prokaryotes is mRNA transport. Because eukaryotic transcription and translation is compartmentally separated, eukaryotic mRNAs must be exported from the nucleus to the cytoplasm. Mature mRNAs are recognized by their processed modifications and then exported through the nuclear pore.

TRANSLATION

Because prokaryotic mRNA does not need to be processed or transported, translation by the ribosome can begin immediately after the end of transcription. Therefore, it can be said that prokaryotic translation is *coupled* to transcription and occurs *co-transcriptionally*.

Eukaryotic mRNA that has been processed and transported to the cytoplasm (i.e. mature mRNA) can then be translated by the ribosome. Translation may occur at ribosomes free-floating in the cytoplasm, or directed to the endoplasmic reticulum by the signal recognition particle. Therefore, unlike prokaryotes, eukaryotic translation *is not* directly coupled to transcription.

DEGRADATION

After a certain amount of time, the message is degraded by RNases. The limited lifetime of mRNA enables a cell to alter protein synthesis rapidly in response to its changing needs. Different mRNAs within the same cell have distinct lifetimes (stabilities). In bacterial cells, individual mRNAs can survive from seconds to more than an hour; in mammalian cells, mRNA lifetimes range from several minutes to days. The greater the stability of an mRNA, the more protein may be produced from that mRNA.

The presence of AU-rich elements in some mammalian mRNAs tends to destabilize those transcripts through the action of cellular proteins that bind these motifs. Rapid mRNA degradation via AU-rich elements is a critical mechanism for preventing the overproduction of potent cytokines such as tumor necrosis factor (TNF) and granulocyte-macrophage colony stimulating factor (GM-CSF). Base pairing with a small interfering RNA (siRNA) or microRNA (miRNA) can also accelerate mRNA degradation.

mRNA STRUCTURE

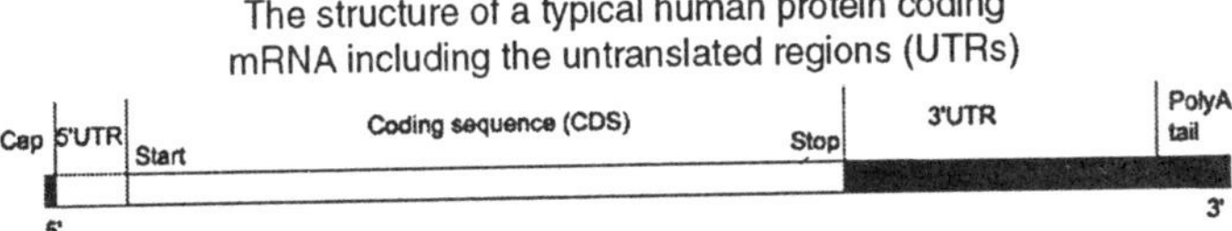

Fig. The Structure of a Mature Eukaryotic mRNA. A Fully Processed mRNA Includes a 5' cap, 5' UTR, Coding Region, 3' UTR, and Poly(A) tail.

5' Cap

The *5′ cap* is a modified guanine nucleotide added to the "front" (5' end) of the pre-mRNA using a 5',5-Triphosphate linkage. This modification is critical for recognition and proper attachment of mRNA to the ribosome, as well as protection from 5' exonucleases. It may also be important for other essential processes, such as splicing and transport.

Coding Regions

Coding regions are composed of codons, which are decoded and translated into one (mostly eukaryotes) or several (mostly prokaryotes) proteins by the ribosome. Coding regions begin with the start codon and end with the one of three possible stop codons. In addition to protein-coding, portions of coding regions may also serve as regulatory sequences in the pre-mRNA as exonic splicing enhancers or exonic splicing silencers. Start codons are indicated by a AUG triplet. Stop codons are indicated by a UAA, UAG, or UGA.

Untranslated Regions

Untranslated regions (UTRs) are sections of the mRNA before the start codon and after the stop codon that are not translated, termed the five prime untranslated region (5' UTR) and three prime untranslated region (3' UTR), respectively. These regions are transcribed with the coding region and thus are exonic as they are present in the mature mRNA. Several roles in gene expression have been attributed to the untranslated regions, including mRNA stability, mRNA localization, and translational efficiency. The ability of a UTR to perform these functions depends on the sequence of the UTR and can differ between mRNAs.

The stability of mRNAs may be controlled by the 5' UTR and/or 3' UTR due to varying affinity for RNA degrading enzymes called ribonucleases and for ancillary proteins that can promote or inhibit RNA degradation. Translational efficiency, including sometimes the complete inhibition of translation, can be controlled by UTRs. Proteins that bind to either the 3' or 5' UTR may affect translation by influencing the ribosome's ability to bind to the mRNA.

MicroRNAs bound to the 3' UTR also may affect translational efficiency or mRNA stability. Cytoplasmic localization of mRNA is thought to be a function of the 3' UTR. Proteins that are needed in a particular region of the cell can actually be translated there; in such a case, the 3' UTR may contain sequences that allow the transcript to be localized to this region for translation.

Some of the elements contained in untranslated regions form a characteristic secondary structure when transcribed into RNA. These structural mRNA elements are involved in regulating the mRNA. Some, such as the SECIS element, are targets for proteins to bind. One class of mRNA element, the riboswitches, directly bind small molecules, changing their fold to modify levels of transcription or translation. In these cases, the mRNA regulates itself.

3' Poly(A) Tail

The 3' poly(A) tail is a long sequence of adenine nucleotides (often several hundred) added to the "tail" or 3' end of the pre-mRNA through the action of an enzyme, polyadenylate polymerase. In higher eukaryotes, the poly(A) tail is added onto transcripts that contain a specific sequence, the AAUAAA signal. The importance of the AAUAAA signal is demonstrated by a mutation in the human alpha 2-globin gene that changes the original sequence AATAAA into AATAAG, which can lead to hemoglobin deficiencies.

Monocistronic Versus Polycistronic Mrna

An mRNA molecule is said to be monocistronic when it contains the genetic information to translate only a single protein. This is the case for most of the eukaryotic mRNAs. On the other hand, polycistronic mRNA carries the information of several proteins, which are translated into several proteins. Most of the mRNA found in bacteria and archea are polycistronic. Dicistronic is the term used to describe a mRNA that encodes only two proteins.

TRANSFER RNA

Transfer RNA (abbreviated tRNA) is a small RNA (usually about 74-95 nucleotides) that transfers a specific amino acid to a growing polypeptide chain at the ribosomal site of protein synthesis during translation. It has a 3' terminal site for amino acid attachment. This covalent linkage is catalyzed by an aminoacyl tRNA synthetase.

It also contains a three base region called the anticodon that can base pair to the corresponding three base codon region on mRNA. Each type of tRNA molecule can be attached to only one type of amino acid, but because the genetic code contains multiple codons that specify the same amino acid, tRNA molecules bearing different anticodons may also carry the same amino acid.

STRUCTURE

tRNA has primary structure, secondary structure (usually visualized as the *cloverleaf structure*), and tertiary structure (all tRNAs have a similar L-shaped 3D structure that allows them to fit into the P and A sites of the ribosome).

- The 5'-terminal phosphate group.
- The acceptor stem is a 7-bp stem made by the base pairing of the 5'-terminal nucleotide with the 3'-terminal nucleotide (which contains the CCA 3'-terminal group used to attach the amino acid). The

acceptor stem may contain non-Watson-Crick base pairs.

- The CCA tail is a CCA sequence at the 3' end of the tRNA molecule. This sequence is important for the recognition of tRNA by enzymes critical in translation. In prokaryotes, the CCA sequence is transcribed. In eukaryotes, the CCA sequence is added during processing and therefore does not appear in the tRNA gene.
- The D arm is a 4 bp stem ending in a loop that often contains dihydrouridine.
- The anticodon arm is a 5-bp stem whose loop contains the anticodon.
- The T arm is a 5 bp stem containing the sequence TØC where Ø is a pseudouridine.
- Bases that have been modified, especially by methylation, occur in several positions outside the anticodon. The first anticodon base is sometimes modified to inosine (derived from adenine) or pseudouridine (derived from uracil).

ANTICODON

An anticodon is a unit made up of three nucleotides that correspond to the three bases of the codon on the mRNA. Each tRNA contains a specific anticodon triplet sequence that can base-pair to one or more codons for an amino acid. For example, one codon for lysine is AAA; the anticodon of a lysine tRNA might be UUU. Some anticodons can pair with more than one codon due to a phenomenon known as wobble base pairing.

Frequently, the first nucleotide of the anticodon is one of two not found on mRNA: inosine and pseudouridine, which can hydrogen bond to more than one base in the corresponding codon position. In the genetic code, it is common for a single amino acid to be specified by all four third-position possibilities; for example, the amino acid glycine is coded for by the codon sequences GGU, GGC, GGA, and GGG.

To provide a one-to-one correspondence between tRNA

molecules and codons that specify amino acids, 61 tRNA molecules would be required per cell. However, many cells contain fewer than 61 types of tRNAs because the wobble base is capable of binding to several, though not necessarily all, of the codons that specify a particular amino acid.

AMINOACYLATION

Aminoacylation is the process of adding an aminoacyl group to a compound. It produces tRNA molecules with their CCA 3' ends covalently linked to an amino acid. Each tRNA is aminoacylated (or *charged*) with a specific amino acid by an aminoacyl tRNA synthetase. There is normally a single aminoacyl tRNA synthetase for each amino acid, despite the fact that there can be more than one tRNA, and more than one anticodon, for an amino acid. Recognition of the appropriate tRNA by the synthetases is not mediated solely by the anticodon, and the acceptor stem often plays a prominent role.

Reaction:

- Amino acid + ATP → aminoacyl-AMP + PPi
- Aminoacyl-AMP + tRNA → aminoacyl-tRNA + AMP

BINDING TO RIBOSOME

The ribosome has three binding sites for tRNA molecules: the A, P and E sites. During translation the A site binds an incoming aminoacyl-tRNA as directed by the codon currently occupying this site. This codon specifies the next amino acid to be added to the growing peptide chain. The A site only works after the first aminoacyl-tRNA has attached to the P site. The P-site codon is occupied by peptdyl-tRNA that is a tRNA with multiple amino acids attached as a long chain. The P site is actually the first to bind to aminoacyl tRNA. This tRNA in the P site carries the chain of amino acids that has already been synthesized. The E site is occupied by the empty tRNA as it is about to exit the ribosome.

tRNA GENES

Organisms vary in the number of tRNA genes in their genome. The nematode worm C. *elegans,* a commonly used model organism in genetics studies, has 29,647 genes in its nuclear genome, of which 620 code for tRNA. The budding yeast *Saccharomyces cerevisiae* has 275 tRNA genes in its genome. In the human genome, which according to current estimates has about 27,161 genes in total, there are about 4,421 non-coding RNA genes, which include tRNA genes. There are 22 mitochondrial tRNA genes; 497 nuclear genes encoding cytoplasmic tRNA molecules and there are 324 tRNA-derived putative pseudogenes.

Cytoplasmic tRNA genes can be grouped into 49 families according to their anticodon features. These genes are found on all chromosomes, except 22 and Y chromosome. High clustering on 6p is observed (140 tRNA genes), as well on 1 chromosome. tRNA molecules are transcribed (in eukaryotic cells) by RNA polymerase III, unlike messenger RNA which is transcribed by RNA polymerase II. pre-tRNAs contain introns; in bacteria these self-splice, whereas in eukaryotes and archaea they are removed by tRNA splicing endonuclease.

The existence of tRNA was first hypothesized by Francis Crick, based on the assumption that there must exist an adapter molecule capable of mediating the translation of the RNA alphabet into the protein alphabet. Significant research on structure was conducted in the early 1960s by Alex Rich and Don Caspar, two researchers in Boston, the Jacques Fresco group in Princeton University and a United Kingdom group at King's College London.

A later publication reported the primary structure in 1965 by Robert W. Holley. The secondary and tertiary structures were derived from X-ray crystallography studies reported independently in 1974 by American and British research groups headed, respectively, by Alexander Rich and Aaron Klug.

FUNCTIONAL ASPECTS OF MODIFIED BASES IN tRNA

Transfer RNA molecules (tRNAs) are typically about 75 nucleotides long and fold into stable tertiary structures not unlike polypeptides. While they have no independent chemical function, their structures are finely tuned to suit a number of steps in translation. They must be specifically recognized by the enzymes that attach amino acids to them, they must bind efficiently to the catalytic sites in the ribosome and they must form productive interactions with mRNA transcripts.

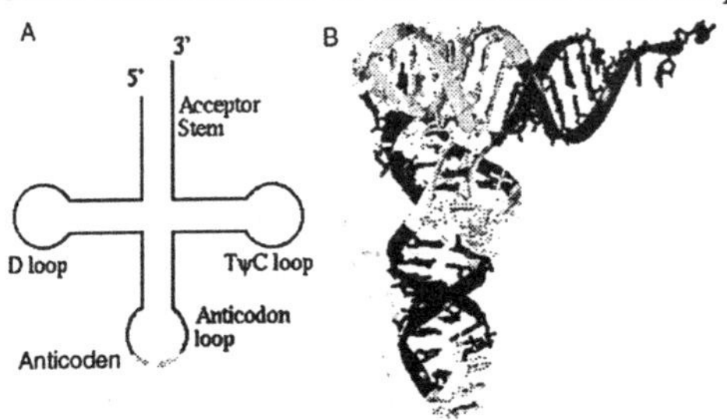

Fig. Two Representations of the Structure of tRNA

The focus of this exercise will be on the adaptation of tRNA structure that permits efficient and specific interactions with mRNA. The region of the tRNA molecule that interacts with the messenger is known as the anticodon. It is comprised of three nucleotides that form part of the anticodon loop. These three bases are free to hydrogen bond with other nucleotides, particularly a set of three base sequences (codons) on mRNA that correspond to the amino acid for which the tRNA molecule is specific. In principle, all three bases of the anticodon could interact via normal Watson- Crick base-pairing to form a three base pair duplex with a single codon sequence.

In fact, the first nucleotide of the anticodon, position 34 in tRNA(Phe), which pairs with the third base of the codon, is known as the "wobble" base. The degeneracy of the genetic code means that several codons can specify a single amino acid. To reduce the number of tRNAs necessary to read the codons, the wobble position is often intended to base pair with two or three different bases at the 3' end of the codon. This reduces the specificity of interaction between the tRNA molecule and the transcript, and has the potential to lead to significant errors in translation if only two base pairs provide the thermodynamic stability to determine the interaction.

It appears that one approach taken by nature to enhance

the specific base pairing interactions between tRNA and mRNA is to use modified bases at positions flanking the anticodon in the tRNA molecule. tRNA is notable for the number and variety of modified bases in its structure.

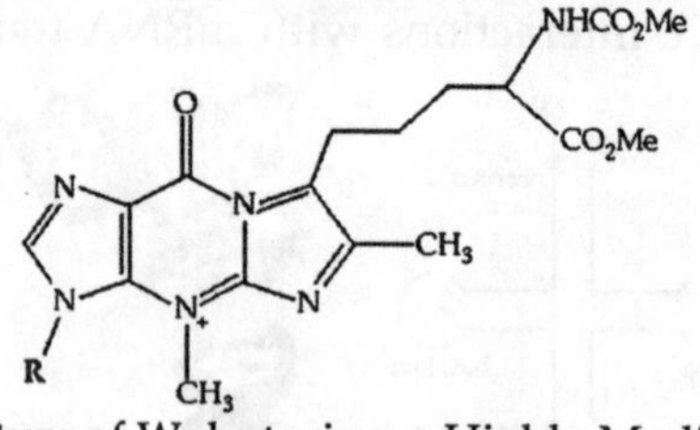

Fig. The Structure of Wybutosine, a Highly Modified Guanine Analog that Appears at Position 37, 3' to the Anticodon, in Yeast tRNA(Phe).

These bases are typically modified forms of the usual A, U, C and G's. Some are methylated analogues (for example 5-methyl uracil - otherwise known as thymine - is found in tRNA), and others are more dramatic, such as wybutosine. These modifications are post-transcriptional. They are made after the tRNA molecule is transcribed from its gene.

The purpose of this exercise is to investigate how modifications made to tRNA(Phe) from yeast, including the creation of wybutosine adjacent to the anticodon, aid in the function of that molecule. tRNA(Phe) has an anticodon sequence GAA, that binds to sequences UUC and UUU on the messenger (both codons for phenylalanine). Because of the wobble at position 3 of the codon, and the U:A base pairs at positions 1 and 2, the interaction between tRNA(Phe) and mRNA is weaker than most. Here, we will examine how this potential problem is overcome.

THE MODEL

Three models are required for this exercise, though a number of others will need to be created in the course of the work. The first comes from coordinates for tRNA(Phe) deposited to the PDB by Kim and co-workers.*(1)* 6tna.pdb is a slightly modified version of the actual PDB entry, with a few changes made for ease of manipulation in Midas. The The second model (arna.pdb is a undecamer duplex of RNA, derived from idealized coordinates of RNA in the A conformation, containing the sequence UUC at the 5' end of one of the strands. brna.pdb is a decamer of duplex RNA, like rna.pdb, but in the B conformation.

RIBOSOMAL rNA

Ribosomal RNA (rRNA) is the central component of the ribosome, the protein manufacturing machinery of all living cells. The function of the rRNA is to provide a mechanism for decoding mRNA into amino acids and to interact with the tRNAs during translation by providing peptidyl transferase activity.

INSIDE THE RIBOSOME

The ribosome is composed of two subunits, named for how rapidly they sediment when subject to centrifugation. tRNA is sandwiched between the small and large subunits and the ribosome catalyzes the formation of a peptide bond between the 2 amino acids that are contained in the tRNA.

The ribosome also has 3 binding sites called A, P, and E.

- The A site in the ribosome binds to an aminoacyl-tRNA (a tRNA bound to an amino acid).
- The NH_2 group of the aminoacyl-tRNA which contains the new amino acid, attacks the carboxyl group of peptidyl-tRNA (contained within the P site) which contains the last amino acid of the growing chain called peptidyl transferase reaction.
- The tRNA that was holding on the last amino acid is moved to the E site, and what used to be the aminoacyl-tRNA is now the peptidyl-tRNA.

A single mRNA can be translated simultaneously by multiple ribosomes.

PROKARYOTES VS. EUKARYOTES

Both prokaryotic and eukaryotic can be broken down into two subunits (the S in 16S represents Svedberg units):

Type	Size	Large Subunit	Small Subunit
prokaryotic	70S	50S (5S, 23S)	30S (16S)
eukaryotic	80S	60S (5S, 5.8S, 28S)	40S (18S)

Note that the S units of the subunits cannot simply be added because they represent measures of sedimentation rate rather than of mass. The sedimentation rate of each subunit is affected by its shape, as well as by its mass.

Prokaryotes

In prokaryotes a small 30S ribosomal subunit contains the 16S rRNA. The large 50S ribosomal subunit contains two rRNA species (the 5S and 23S rRNAs). Bacterial 16S, 23S, and 5S rRNA genes are typically organized as a co-transcribed operon. There may be one or more copies of the operon dispersed in the genome (for example, *Escherichia coli* has seven).

Archaea contains either a single rDNA operon or multiple copies of the operon. The 3' end of the 16S rRNA (in a ribosome) binds to a sequence on the 5' end of mRNA called the Shine-Dalgarno sequence.

Eukaryotes

In contrast, eukaryotes generally have many copies of the rRNA genes organized in tandem repeats; in humans approximately 300–400 rDNA repeats are present in five clusters (on chromosomes 13, 14, 15, 21 and 22).

The 18S rRNA in most eukaryotes is in the small ribosomal subunit, and the large subunit contains three rRNA species (the 5S, 5.8S and 28S rRNAs).

Mammalian cells have 2 mitochondrial (12S and 16S) rRNA molecules and 4 types of cytoplasmic rRNA (28S, 5.8S, 5S (large ribosome subunit) and 18S (small subunit). 28S, 5.8S, and 18S rRNAs are encoded by a single transcription unit (45S) separated by 2 Internally transcribed spacer (ITS). The 45S rDNA organized into 5 clusters (each has 30-40 repeats) on chromosomes 13, 14, 15, 21, and 22. These are transcribed by RNA polymerase I. 5S occurs in tandem arrays (~200-300 true 5S genes and many dispersed pseudogenes), the largest one on the chromosome 1q41-42. 5S rRNA is transcribed by RNA polymerase III.

The tertiary structure of the small subunit ribosomal RNA (SSU rRNA) has been resolved by X-ray crystallography. The secondary structure of SSU rRNA contains 4 distinct domains — the 5', central, 3' major and 3' minor domains. A model of the secondary structure for the 5' domain (500-800 nucleotides) is shown.

Translation

Translation is the net effect of proteins being synthesized by ribosomes, from a copy (mRNA) of the DNA template in the nucleus. One of the components of the ribosome (16s rRNA) base pairs complementary to a sequence upstream of the start codon in mRNA.

IMPORTANCE OF rRNA

Ribosomal RNA characteristics are important in medicine and in evolution.

- rRNA is the target of several clinically relevant antibiotics: chloramphenicol, erythromycin, kasugamycin, micrococcin, paromomycin, ricin, sarcin, spectinomycin, streptomycin, and thiostrepton.
- rRNA is the most conserved (least variable) gene in all cells. For this reason, genes that encode the rRNA (rDNA) are sequenced to identify an organism's taxonomic group, calculate related groups, and estimate rates of species divergence. For this reason many thousands of rRNA sequences are known and stored in specialized databases such as RDP-II and the European SSU database.

SYNTHESIS OF RIBOSOMAL RNA - rRNA

The 70S ribosome of prokaryotes consists of a 308 subunit and a 50S subunit. The former contains 16S rRNA and the latter 238 and 58 rRNAs. In bacterial genes the sequences specifying 168, 238 and some times also5S RNA are arranged in a series mRNA is transcribed from DNA as a 30S unit, which bas been called p308. Experimental evidence indicates that the 308 unit has the 16S component at the 5' end and the 238 component at the 3' end, with spacer units between the two components. In some prokaryotes 5S rRNA is transcribed at or near the 3' end.

During processing the p30S transcriptional unit is cleaved within the spacer segment by RNase III into 25S and l8S segments. These are then reduced to p23S and pl6S segments respectively, also by RNase III. Secondary trimming of these intermediates yields the final size, 23S and 168, respectively.

The enzyme involved in secondary trimming is probably a ribonuclease, designated as RNase M or maturase.In E. coli trimming of the pl6s component involves removal of a total of about 200 nucleotides from both 3' and 5' sides. The final cleavages occur when the rRNAs are associated with structural proteins of ribosomes in the 'preribosomal particles'.

Some nucleoside modifications take place in the p30S component, while others occur only after cleavage. Methylaiions found in 23S rRNA occur early at the p30S stage, whereas modifications found in 168 rRNA take place at a late stage of processing, in some cases after association with specific ribosomal proteins.

In E. coli the precursors of 58 rRNA are only a few nucleotides larger than mature SS rRNA. In Bacillus there are two types of larger precursors. One, designated as p5A, is 180 nucleotides long, while the other, P5B' is 140-150 nucleotides long. Both are apparently transcribed on different 5S genes.

Cleavage of p5A by the enzyme RNase M5 splits it into three components, a 5' terminal piece of about 20 nucleotides, mature 5S rRNA of 118 nucleotides and a 3' terminal piece of about 40 nucleotides. The 5' and 3' terminal pieces are broken down to their monouculeotides by an exonuclease, which thus serves a scavenging function.

Mature 5S rRNA is resistant to this enzyme. The precursor p5B is also split by RNase M5 into a 22 nucleotide 5' terminal piece, a mature 58 mRNA and some smaller 3' terminal pieces. In eukaryotes the 808 ribosome consists of 40S and 60S subunits. The 408 subunits contain 16-18S rRNA, and the 60S subunits 25-28S rRNA, S.88 RNAand.5S rRNA.

The genes coding for 16-188 rRNA and 25S-288 rRNA arc arranged in clusters of hundreds to thousands of copies. In eukaryotes the transcribed rRNA is 45S rRNA in mammals and 36-38S rRNA in lower eukaryotes, with molecular weights of 4.5 million and 2.6-2.8 million, respectively. There are currently two schemes for the possible arrangement of 28S and 18S segments within the 458 rRNA precursor; and the processing of the precursor rRNA.

According to one scheme 458 rRNA has the general form: 5′P-28s rRNA - spacer - 18SRNA-spacer-3' OH. This transcriptional RNA has a life time of about 15 minutes during which methylation of the ribose moiety in the 288 and 188 regions takes place. An endonuclease cleaves 458 rRNA into 418 and 208components.

The cleavage takes place towards the 5' side of the 208 rRNA components. 41S rNA is degraded through 36S and 328 stages to mature 28S rRNA. During the three steps cleavage takes place in the non-methylated spacer segment by exonucleases. Similarly degration of the non methylated spacer region of the 20S component yields 18S rRNA.

According to the second scheme the transcribed spacer sequence is situated at the 5′P end of 45S rRNA and the 28S com-ponent at the 3′OH end. The bulk of the experimental evidence supports this scheme which would also unify transcription in prokaryotes and eukaryotes. Processing occurs almost entirely within the nucleolus in eukaryote cells. The number of nucleoli within a cell depends on the extent to which the rRNA genes within a chromosome are dispersed.

45S transcriptional rRNA undergoes cleavage at four sites. Cleavage at site I removes the transcribe4 spacer sequence at the 5′P end. In primitive eukaryotes it may occur before transcription is completed. The order of cleavage at sites 2 and 3 varies in different species. Cleavage at site 2 separates 18S rRNA.

A segment of 140 nucleotides, with a sedimentation coefficient of 5.8S (formerly 78), situated between sites 3 and 4, remains covalently linked to the segment formed by.cleavage 3. In yeast tRNAs about 10% of 5.88 RNA is extended at the 5' end by 6-7 nucleotides. Cleavage 4 results in final trimming of the large rRNA segment. 58 rRNA is transcribed separately.

Analyses of the terminal sequences of 16-18S rRNA from

many organisms reveal that the 3' end is almost always A-OH and the 5' end is pU. As many as 8 nucleotides at the 3' end are apparently uniform. They may be involved in the binding of mature rRNA to mRNA. Similarly 6 or more nucleotides may be uniform at the 5' end.

Nucleoside modifications take place on the initial transcript 45S rRNA in regions that give rise to mature 285 and 18S rRNAS. A few base methylations (6 in 18S rRNA) occur at later stages. This late methylation is characteristic of prokaryotes. Mefhylation seems to be essential for cleavage.

CLOSURE OF rRNA

Bacterial genome projects usually comprise three well-defined stages. The first one is the construction of high quality random shotgun libraries with varying insert sizes. Following this, the next task is typically to generate a deep coverage of the genome, with random shotgun sequences. This step generates the bulk of the sequences that are used to compile the finished genome sequence, and it is in many ways the heart of the project. Shotgun sequences are then assembled into sequence contigs, each representing a separate, non-overlapping portion of the genome.

The resulting assembly is then converted into a continuous genome sequence through a more laborious and time-consuming finishing phase. Due to the small size and low complexity of bacterial genomes, a genome sequence is only considered finished when all bases meet an acceptable quality level and the individual sequences are assembled into a single continuous consensus sequence. During the finishing phase of a bacterial genome project, the overall sequence quality is improved and sequence data are generated to close gaps between existing contigs. Finishing can take a variety of forms, but it almost always involves the generation of sequences from large insert clones that have been demonstrated to span virtual gaps in the shotgun assembly, and by the use of alternative strategies to close real gaps when the corresponding DNA fragment has not appeared in any of the libraries generated for the sequencing project.

The occurrence of real gaps is often associated with statistical cloning fluctuation of the shotgun model, repetitive regions in the genome (for example ribosomal operons), and with sequences refractory to the cloning system used. Strategies for real gap closure that have been used so far include the use of alternative cloning vectors, combinatory PCR, with primers derived from contig ends, subtractive hybridization, physical mapping, and direct sequencing of genomic DNA. The time required for finishing depends on the nature of the genome and most particularly on the structure and frequency of repetitive sequences, but this phase easily outlasts that of the generation of the shotgun sequences and is indeed one of the key rate-limiting steps in bacterial genome sequencing.

Here we describe the assembly process and finishing phase of the *Chromobacterium violaceum* genome project. *C. violaceum* is a free-living, gram-negative bacteria that is highly abundant in the water and banks of the Negro River in the Brazilian Amazon, and produces a violacein pigment with antimicrobial activity against some important pathogens as well as antiviral and anticancer activity. The sequencing and analysis of the *C. violaceum* genome were entirely executed by the Brazilian National Genome Sequencing Consortium. A random shotgun strategy was used, and the resulting sequences were assembled into 57 contigs.

These were then organized into 19 scaffolds, using the information from shotgun and cosmid clones; sequences of which were located in different contigs. Forty-seven virtual gaps within the scaffolds were closed by whole insert sequencing of the corresponding shotgun/cosmid clones. Eighteen real gaps were, for the most part, closed by applying the PCR-assisted contig extension (PACE), followed by specific confirmatory and oriented combinatory PCR. PACE involves the generation of stepwise extensions from the ends of contigs by PCR, until the closure of individual gaps is achieved. This methodology has proven to be especially useful for extending the multicopy ribosomal operons, and has greatly accelerated the finishing phase of the *C. violaceum* genome project.

METHODS

Pace Methodology

PACE was developed as a two-step PCR strategy, with nested primers derived from contig ends, as described by Carraro et al., 2003. Briefly, a set of 96 primers was randomly selected, with no reference to their precise sequence. Pairs of outward-facing specific nested primers were then designed, approximately 140 bp from contig ends and 40 bp apart from each other. Specific primers were checked for alignment to a single position of the genome with the FASTA programme. Secondary structures and dimer formation were verified using the Oligo Tech software. PCR was then performed in 96-well plates, using 80 ng genomic DNA as templates for the first reaction and 1 ìl of a 1:100 dilution of the first reaction as templates in a subsequent nested reaction.

Specific PCR

All contig extensions were confirmed by specific PCR, followed by direct sequencing of the resulting PCR products. Reaction mixtures for specific PCR contained 80 ng genomic DNA, 250 ìM dNTP, 1.5 mM MgC12, 1 U Platinum High Fidelity *Taq* DNA polymerase (Invitrogen, Carlsbad, CA, USA), and 10 ìM of each specific primer in a final volume of 20 ìl. Reactions were carried out at 94°C for 30 s, 60°C for 30 s, and 68°C for 2 min from 35 cycles. An initial cycle of 94°C for 2 min, and a final extension at 68°C for 6 min was used.

Combinatory PCR

For ribosomal operon orientation, primers were designed in the flanking regions of the 16S and 5S rRNA genes (outside of the operon repeat unit) and PCRs were undertaken in a combinatory way. The expected fragment sizes were between 5.2 and 5.8 kb, depending on the annealing position of the primers in the flanking region. PCRs were performed in a final volume of 20 ìl, containing 1.5 mM MgC12, 300 ìM dNTPs, 2 U Platinum High Fidelity *Taq* DNA polymerase (Invitrogen) and 15 ìM of each flanking primer.

Reactions were carried out at 94°C for 30 s, 60°C for 30 s and 68°C for 5 min, for 35 cycles. Initial denaturation step at 94°C for 2 min, and final extension step at 68°C for 6 min, was used. Two almost identical versions of the rRNA operon were identified during the assembly phase. The difference between the two copies resides in a 100-bp insertion in the intergenic region between the 16S and ILE genes and a 74-bp insertion between the ILE and ALA genes (copy A - contains 100 bp-16S/ILE, 74 bp- ILE/ALA and copy B - does not contain 100 bp-16S/ILE, 74 bp-ILE/ALA).

In order to position correctly the two versions in the genome, a primer located 150 bp downstream of the 100-bp insertion (in the ILE gene) was used in a combinatory way in amplification reactions, together with one of the eight specific primers corresponding to the 16S rRNA flanking region. PCR reactions were carried out as described above, except for modifications in the cycling parameters (94°C for 30 s, 60°C for 30 s and 68°C for 2 min). Sequence specificity was checked by BLASTN analysis of the two fragment extremities against the two available ribosomal operon copies. A fragment was considered specific to one of the two copies if the high quality sequence portion aligned with the corresponding copy with at least 95% identity.

Product Analysis

Three to five microliters from each PCR were loaded onto a 1% ethidium bromidestained agarose gel. Single PACE products, or fragments of expected size in the case of confirmatory or combinatory PCRs, were purified with the QIAquickTM PCR Purification Kit and sequenced directly using the same specific primer used for amplification on an ABI PrismR 3100 DNA sequencer. High-quality sequences with more than 300 bp, and Phred quality greater than 20, were analyzed for specificity. Sequence specificity was checked by BLASTN analysis against the available genome assembly, and a fragment was considered specific if at least 25 bp of the high quality sequence aligned with at least 95% identity with the end of the corresponding contig.

Chromobacterium Violaceum Genome Assembly

The sequencing and analysis of the *C. violaceum* genome were entirely executed by the Brazilian National Genome Sequencing Consortium, comprising 25 sequencing laboratories, one bioinformatics centre, and three coordination laboratories, distributed throughout Brazil. In the initial phase of the project, random shotgun sequencing produced approximately 80,000 high quality reads, generated from both ends of pUC18 clones, with insert sizes raging from 2.0 to 4.0 kb. Additionally, both ends of 3,350 cosmid clones, with an average insert size of 40 kb, were also sequenced, providing a validation check of the final assembly.

The shotgun sequences, corresponding to approximately 13-fold genome coverage, were assembled into 57 contigs. These shotgun contigs were then organized into 19 scaffolds, using the information from shotgun and cosmid clones, the end sequences of which were located in different contigs. Forty-seven virtual gaps within the 19 scaffolds were closed by whole insert sequencing of the corresponding shotgun/cosmid clones.

Points of genome assembly instability and regions of low quality sequences were identified using the Autofinisher Programme, and were resolved by re-sequencing of the selected clones. Real gaps that did not involve the rRNA gene (n = 18) were mainly closed by applying PACE, as detailed in Carraro et al. Here, we relate the methodology used to close rRNA-related gaps and to correctly position the eight ribosomal operon units in the *C. violaceum* genome assembly.

Closure of Real Gaps by Pace

The PACE technique depends on rare mismatched primer-template interactions that occur between arbitrary primers and template DNA with an unknown sequence, even under highly stringent conditions. These are captured through elevated PCR-cycle repetition, and through the use of specific anchoring primers, corresponding to adjacent regions of known sequence (contig ends).

Thus, PACE allows the generation of stepwise extensions

from the ends of contigs by PCR, until the closure of individual gaps is achieved. When we started using PACE to close the *C. violaceum* genome assembly, seven of the existing 38 contig ends ended with the 5S rRNA sequence and three ended with the 16S rRNA sequence, suggesting the existence of at least seven identical copies of the rRNA operon.

Twenty-two PACE reactions were initially applied to extend contig ends that did not contain rRNA-derived sequences, resulting in 137 specific sequences, allowing the immediate closure of 15 real gaps in the *C. violaceum* assembly due to their relatively small size. Of these, six apparently linked the contig to an rRNA gene (one to a 5S rRNA gene and five to 16S rRNA genes), suggesting a total of eight copies of the ribosomal operon.

In two cases (gap 191-221 of 1691 bp, gap 202-199L of 1790 bp), additional rounds of PACE were undertaken until the complete closure of each gap was achieved. All contig extensions and gap closures were confirmed by specific PCR, followed by direct sequencing of the PCR products.

Closure of Gaps Related to rRNA Sequences

We used a single pair of nested primers specific for the 16S rRNA gene to check for the extension of contigs ending in 16S rRNA. Thirty-five PACE fragments were selected for sequencing and high quality sequences were further analyzed using BLASTN to check for their specificity. Of these, 34 were found to be specific, and they confirmed five novel flanking regions for the 16S rRNA gene (contigs 194, 205, 222, 230 and 183), in addition to the three known, giving a total of eight rRNA operons.

All 16S and 5S rRNA contig extensions were confirmed by specific PCR, followed by direct sequencing of the PCR products. Seventy-two specific PCR reactions, including real gaps and rRNA-related gaps, were performed, corresponding to the four possible combinations between the external and nested primers from both contig ends. At least one fragment from each junction was submitted for sequencing to check the specificity.

Chromobacterium Violaceum Genome

After the 16S PACE protocol, the genome assembly was composed of eight unoriented contigs, ending either with 16S rRNA or 5S rRNA sequences. To assemble the contigs in the correct order, we used a combinatory PCR strategy, with primers derived from the non-repetitive genomic region flanking the 16S and 5S rRNA gene. PCR fragments of expected size were sequenced from both extremities, and high quality sequences were aligned to the genome assembly.

The orientation of the contigs was confirmed if at least 100 bp of the sequenced fragment aligned with the end of the corresponding contigs, outside the common ribosomal operon sequence. The whole genome assembly revealed a slight difference between the sequences, corresponding to the ribosomal operons. Two almost identical versions were identified.

The difference between the two versions resided in a 100-bp insertion between the 16S and ILE genes, and 74 bp between the ILE and ALA genes. In order to correctly place the two versions (A and B), in the eight already-oriented ribosomal operon copies, a primer located 150 bp downstream of the 100-bp insertion was used in a combinatory way, together with one of the eight specific primers corresponding to the 16S rRNA-flanking region. Using this strategy, we found out that most (6/8) of the rRNA operon copies contained the 100-bp and 74-bp insertion (copy A). These results were independently confirmed by the finding of a larger number of reads in the genome assembly in which the 100-bp insertion was not present. After the initial assembly of shotgun reads, the *C. violaceum* genome sequence was organized into 19 scaffolds. The genome sequence assembly was mainly made difficult by the eight almost identical rRNA operon copies dispersed throughout the genome.

Using PACE, we were able to generate contig extensions with an average of 1 kb in length from all contigs, which closed the majority of gaps in a single round of experimentation. In addition, the PACE methodology proved to be extremely useful for extending the multicopy ribosomal operons. The surprisingly high success rate of our approach can be attributed to deep shotgun coverage and the small size of the gaps in the *C. violaceum* genome assembly. However, deep shotgun coverage is not a pre-requisite for using pace.

Analyses made by our group demonstrated that the methodology can be applied at early stages of the bacterial genome assembly, drastically reducing the time and cost of the finishing phase. Usually the finishing phase of a genome project takes between 50 to 60% of the total time of the project. In the case of the *C. violaceum* genome project we used approximately 30% of the total time required to conclude the project, corresponding to a reduction of at least 40% of the estimated time.

The importance of PACE is that, at a pre-determined point in the shotgun sequencing, primers can be generated and contig extensions obtained with no requirement for a one-by-one analysis of the gaps. Using successive PACE reactions, it is possible to close gaps in bacterial genomes in a stepwise fashion, regardless of their size. Based on the experience accumulated in the *C. violaceum* genome project we strongly recommend the use of the PACE methodology in the finishing phase of bacterial genome projects.

Chapter 9

Protein Synthesis

ONE-GENE-ONE-PROTEIN

During the 1930s, despite great advances, geneticists had several frustrating questions yet to answer:

What exactly are genes?

How do they work?

What produces the unique phenotype associated with a specific allele?

Answers from physics, chemistry, and the study of infectious disease gave rise to the field of molecular biology. Biochemical reactions are controlled by enzymes, and often are organized into chains of reactions known as metabolic pathways. Loss of activity in a single enzyme can inactivate an entire pathway. Archibald Garrod, in 1902, first proposed the relationship through his study of alkaptonuria and its association with large quantities "alkapton".

He reasoned unaffected individuals metabolized "alkapton" (now called homogentistic acid) to other products so it would not buildup in the urine. Garrod suspected a blockage of the pathway to break this chemical down, and proposed that condition as "an inborn error of metabolism". He also discovered alkaptonuria was inherited as a recessive Mendelian trait. George Beadle and Edward Tatum during the late 1930s and early 1940s established the connection Garrod suspected between genes and metabolism.

They used X rays to cause mutations in strains of the mold *Neurospora*. These mutations affected a single genes and single enzymes in specific metabolic pathways. Beadle and Tatum proposed the "one gene one enzyme hypothesis" for which they won the Nobel Prize in 1958. Since the chemical reactions occurring in the body are mediated by enzymes, and since enzymes are proteins and thus heritable traits, there must be a relationship between the gene and proteins. George Beadle, during the 1940s, proposed that mutant eye colors in *Drosophila* was caused by a change in one protein in a biosynthetic pathway.

In 1941 Beadle and coworker Edward L. Tatum decided to examine step by step the chemical reactions in a pathway. They used *Neurospora crassa* as an experimental organism. It had a short life-cycle and was easily grown. Since it is haploid for much of its life cycle, mutations would be immediately expressed. The meiotic products could be easily inspected. Chromosome mapping studies on the organism facilitated their work.

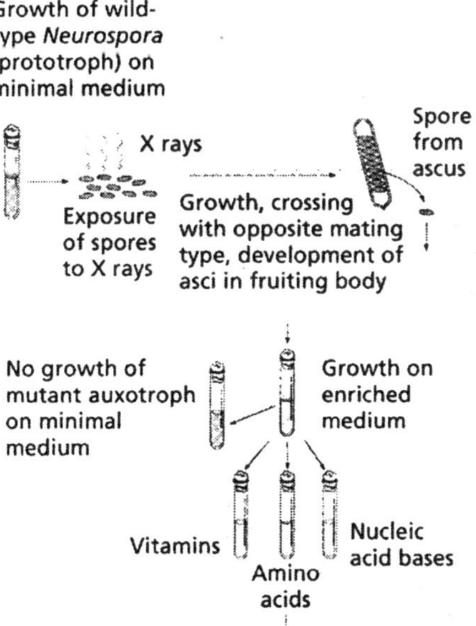

Neurospora can be grown on a minimal medium, and it's nutrition could be studied by its ability to metabolize sugars and other chemicals the scientist could add or delete from the mixture of the medium. It was able to synthesize all of the amino acids and other chemicals needed for it to grow, thus mutants in synthetic pathways would easily show up. X-rays induced mutations in *Neurospora,* and the mutated spores were placed on growth media enriched with all essential amino acids. Crossing the mutated fungi with non-mutated forms produced spores which were then grown on media supplying only one of the 20 essential amino acids.

If a spore lacked the ability to synthesize a particular amino acid, such as Pro (proline), it would only grow if the Proline was in the growth medium. Biosynthesis of amino acids (the building blocks of proteins) is a complex process with many chemical reactions mediated by enzymes, which if mutated would shut down the pathway, resulting in no-growth. Beadle and Tatum proposed the "one gene one enzyme" theory.

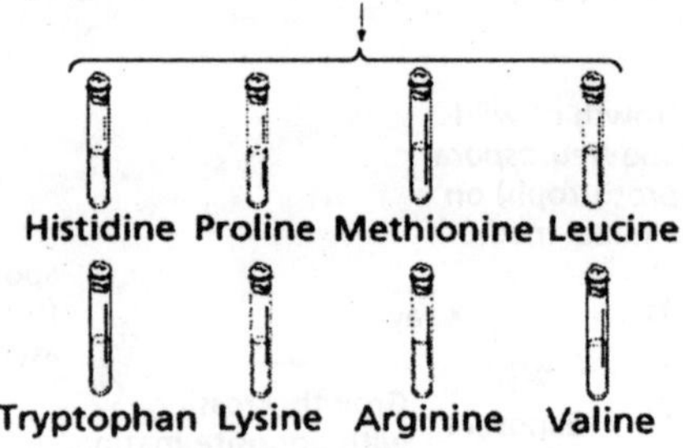

Fig. The Beadle and Tatum Experiment that Suggested the One Gene One Enzyme Hypothesis.

One gene codes for the production of one protein. "One gene one enzyme" has since been modified to "one gene one polypeptide" since many proteins (such as hemoglobin) are made of more than one polypeptide.

THE STRUCTURE OF HEMOGLOBIN

Linus Pauling used electrophoresis to separate hemoglobin molecules. Sickle-cell anemia (h) is a recessive allele in which a defective hemoglobin is made, ultimately causing pain and death to those individuals homozygous recessive for the trait. Pauling reasoned that if Beadle and

Tatum were correct, there should be a slight (but detectable) difference between the structure of a normal (HH) and sickle cell (hh) hemoglobin due to genetic differences. Heterozygotes (Hh, also sampled by Pauling) make both normal and "sickle cell" hemoglobins. Later, Vernon Ingram discovered that the normal and sickle-cell hemoglobins differ by only 1 (out of a total of 300) amino acids.

VIRUSES CONTAIN DNA

The coats of viruses act as antigens, initiating an antigen-specific antibody response. Remember that vaccines work by either prompting the immune system to make antibodies or by supplying antibodies. If a virus (or anything else for that matter) mutates its antigens, the immune system is forever playing catch-up.

RNA LINKS THE INFORMATION IN DNA TO THE SEQUENCE OF AMINO ACIDS IN PROTEIN

Ribonucleic acid (RNA) was discovered after DNA. DNA, with exceptions in chloroplasts and mitochondria, is restricted to the nucleus (in eukaryotes, the nucleoid region in prokaryotes). RNA occurs in the nucleus as well as in the cytoplasm (also remember that it occurs as part of the ribosomes that line the rough endoplasmic reticulum).

Scientists for some time had suspected such a link between DNA and proteins. Cells of developing embryos contain high levels of RNA. Rapidly growing *E. coli* has half its mass as ribosomes. Ribosomes are 2/3 RNA (a type of RNA known as ribosomal RNA or rRNA) and 1/3 protein. RNA is synthesized from viral DNA in an infected cell before protein synthesis begins. Some viruses, for example Tobacco Mosaic Virus (TMV) have RNA in place of DNA. If RNA extracted from a virus was injected into a host cell the cell began to make new viruses. Clearly RNA was involved in protein synthesis.

Crick's central dogma. Information flow (with the exception of reverse transcription) is from DNA to RNA via the process of transcription, and thence to protein via translation. Transcription is the making of an RNA molecule off a DNA template. Translation is the construction of an amino acid sequence (polypeptide) from an RNA molecule. Although originally called dogma, this idea has been tested repeatedly with almost no exceptions to the rule being found (save retroviruses).

Fig. The Central Dogma

The code consists of at least three bases, according to astronomer George Gamow. To code for the 20 essential amino acids a genetic code must consist of at least a 3-base set (triplet) of the 4 bases. If one considers the possibilities of arranging four things 3 at a time (4X4X4), we get 64 possible code words, or codons (a 3-base sequence on the mRNA that codes for either a specific amino acid or a control word).

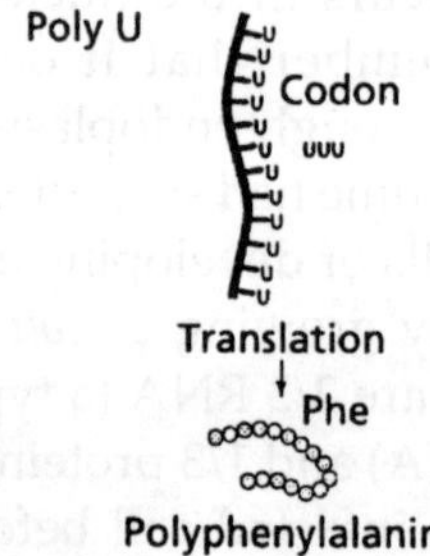

Fig. Steps in Breaking the Genetic Code: the Deciphering of a Poly-U mRNA

The genetic code was broken by Marshall Nirenberg and Heinrich Matthaei, a decade after Watson and Crick's work. Nirenberg discovered that RNA, regardless of its source organism, could initiate protein synthesis when combined with contents of broken E. coli cells. By adding poly-U to each of 20 test-tubes (each tube having a different "tagged" amino acid) Nirenberg and Matthaei were able to determine that the codon UUU (the only one in poly-U) coded for the amino acid phenylalanine.

Likewise, an artificial mRNA consisting of alternating A and C bases would code for alternating amino acids histidine and threonine. Gradually, a complete listing of the genetic code codons was developed.

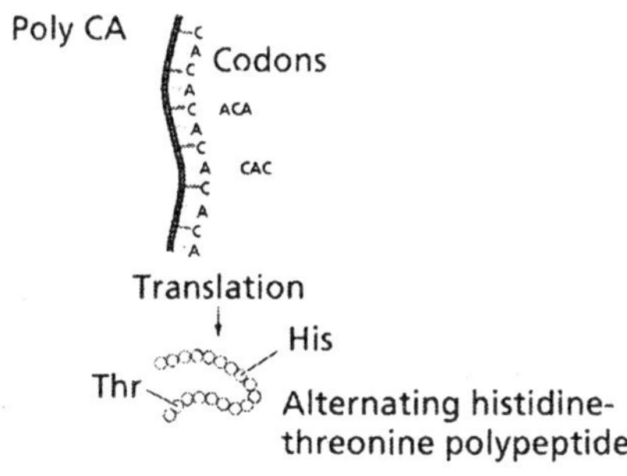

Fig. Deciphering the Code: Poly CA

PROTEIN SYNTHESIS

Prokaryotic gene regulation differs from eukaryotic regulation, but since prokaryotes are much easier to work with, we focus on prokaryotes at this point. Promoters are sequences of DNA that are the start signals for the transcription of mRNA. Terminators are the stop signals. mRNA molecules are long (500- 10,000 nucleotides).

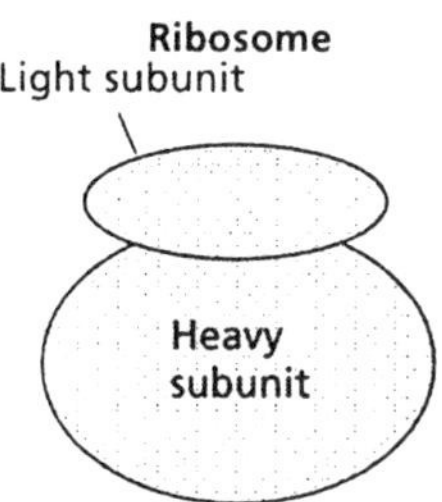

Fig. Subunits of a Ribosome

Ribosomes are the organelle (in all cells) where proteins are synthesized. They consist of two-thirds rRNA and one-third protein. Ribosomes consist of a small (in *E. coli*, 30S) and larger (50S) subunits. The length of rRNA differs in each. The

30S unit has 16S rRNA and 21 different proteins. The 50S subunit consists of 5S and 23S rRNA and 34 different proteins. The smaller subunit has a binding site for the mRNA. The larger subunit has two binding sites for tRNA.

Transfer RNA (tRNA) is basically cloverleaf-shaped. tRNA carries the proper amino acid to the ribosome when the codons call for them. At the top of the large loop are three bases, the anticodon, which is the complement of the codon. There are 61 different tRNAs, each having a different binding site for the amino acid and a different anticodon.

For the codon UUU, the complementary anticodon is AAA. Amino acid linkage to the proper tRNA is controlled by the aminoacyl-tRNA synthetases. Energy for binding the amino acid to tRNA comes from ATP conversion to adenosine monophosphate (AMP).

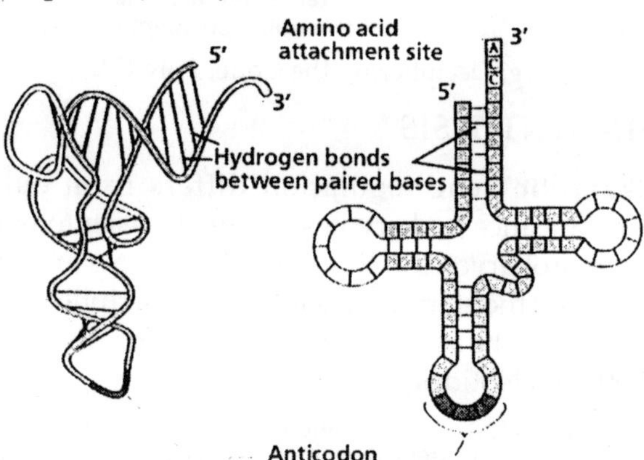

Fig. Two Models of tRNA

Translation is the process of converting the mRNA codon sequences into an amino acid sequence. The initiator codon (AUG) codes for the amino acid N-formylmethionine (f-Met). No transcription occurs without the AUG codon. f-Met is always the first amino acid in a polypeptide chain, although frequently it is removed after translation. The intitator tRNA/ mRNA/small ribosomal unit is called the initiation complex. The larger subunit attaches to the initiation complex. After the initiation phase the message gets longer during the elongation phase.

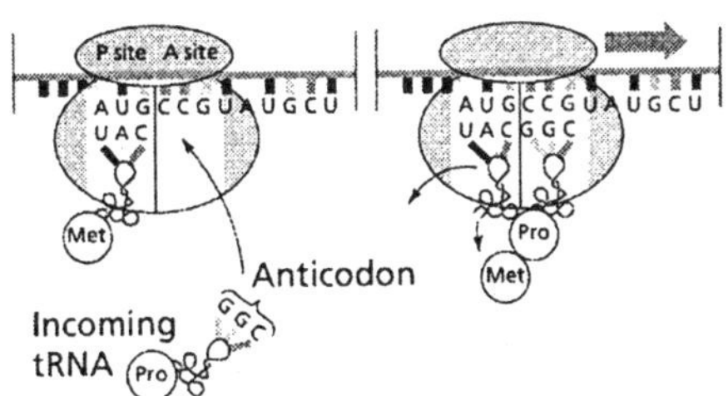

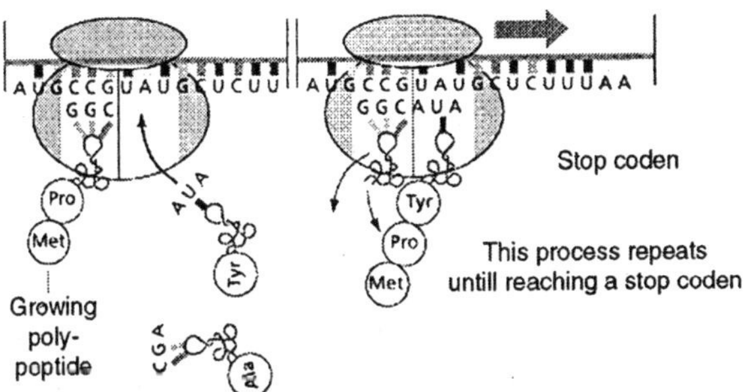

Fig. Translation

New tRNAs bring their amino acids to the open binding site on the ribosome/mRNA complex, forming a peptide bond between the amino acids. The complex then shifts along the mRNA to the next triplet, opening the A site. The new tRNA enters at the A site. When the codon in the A site is a termination codon, a releasing factor binds to the site, stopping translation and releasing the ribosomal complex and mRNA.

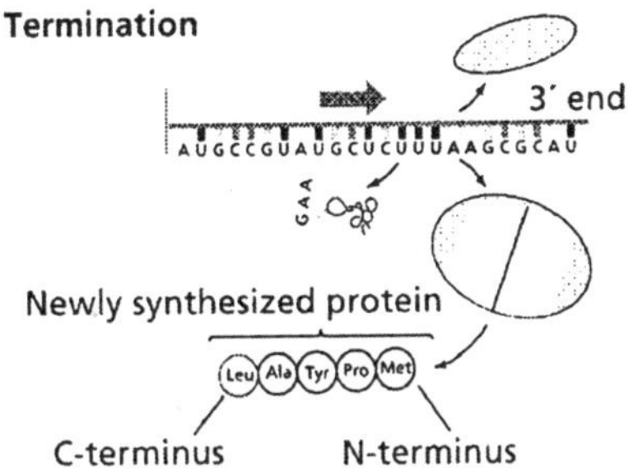

Fig. Termination

Often many ribosomes will read the same message, a

structure known as a polysome forms. In this way a cell may rapidly make many proteins.

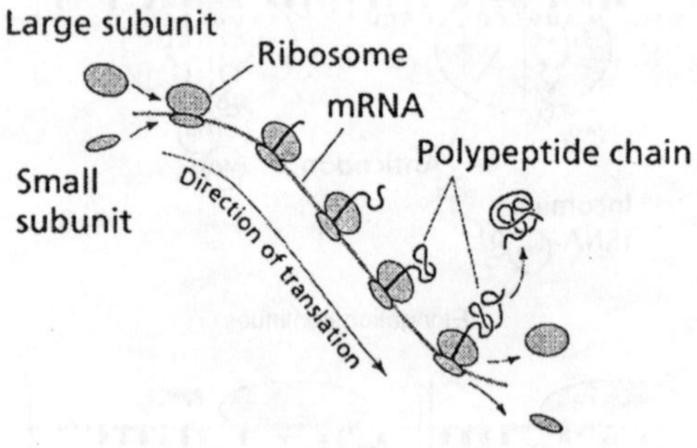

Fig. Many Ribosomes Translating the same Message, a Polysome

MUTATIONS REDEFINED

We earlier defined mutations as any change in the DNA. We now can refine that definition: a mutation is a change in the DNA base sequence that results in a change of amino acid(s) in the polypeptide coded for by that gene. Alleles are alternate sequences of DNA bases (genes), and thus at the molecular level the products of alleles differ (often by only a single amino acid, which can have a ripple effect on an organism by changing).

Addition, deletion, or addition of nucleotides can alter the polypeptide. Point mutations are the result of the substitution of a single base. Frame-shift mutations occur when the reading frame of the gene is shifted by addition or deletion of one or more bases. With the exception of mitochondria, all organisms use the same genetic code. Powerful evidence for the common ancestry of all living things.

PROTEIN BIOSYNTHESIS

Protein biosynthesis (synthesis) is the process in which cells build proteins. The term is sometimes used to refer only to protein translation but more often it refers to a multi-step process, beginning with amino acid synthesis and transcription which are then used for translation. Protein biosynthesis, although very similar, differs between prokaryotes and eukaryotes.

AMINO ACID SYNTHESIS

Amino acids are the monomers which are polymerized to produce proteins. Amino acid synthesis is the set of biochemical processes (metabolic pathways) which build the amino acids from carbon sources like glucose. Not all amino acids may be synthesised by every organism, for example adult humans have to obtain 8 of the 20 amino acids from their diet.

TRANSCRIPTION

Transcription is the process by which an mRNA template, encoding the sequence of the protein in the form of a trinucleotide code, is transcribed from the genome to provide a template for translation. Transcription copies the template from one strand of the DNA double helix, called the template strand. Transcription can be divided into 3 stages: Initiation, Elongation and Termination, each regulated by a large number of proteins such as transcription factors and coactivators that ensure the correct gene is transcribed in response to appropriate signals. The DNA strand is read in the 3' to 5' direction and the mRNA is transcribed in the 5' to 3' direction by the RNA polymerase.

TRANSLATION

The synthesis of proteins is known as translation. Translation occurs in the cytoplasm where the ribosomes are located. Ribosomes are made of a small and large subunit which surrounds the mRNA.

In translation, messenger RNA (mRNA) is decoded to produce a specific polypeptide according to the rules specified by the genetic code. This uses an mRNA sequence as a template to guide the synthesis of a chain of amino acids that form a protein. Translation is necessarily preceded by transcription. Translation proceeds in four phases: activation, initiation, elongation and termination (all describing the growth of the amino acid chain, or polypeptide that is the product of translation). In activation, the correct amino acid (AA) is joined to the correct transfer RNA (tRNA). While this is not technically a step in translation, it is required for translation to proceed. The AA is joined by its carboxyl group to the 3' OH of the tRNA by an ester bond. When the tRNA has an amino acid linked to it, it is termed "charged". Initiation involves the small subunit of the ribosome binding to 5' end of mRNA with the help of initiation factors (IF), other proteins that assist the process.

Elongation occurs when the next aminoacyl-tRNA (charged tRNA) in line binds to the ribosome along with GTP and an elongation factor. Termination of the polypeptide happens when the A site of the ribosome faces a stop codon (UAA, UAG, or UGA). When this happens, no tRNA can recognize it, but releasing factor can recognize nonsense codons and causes the release of the polypeptide chain. The capacity of disabling or inhibiting translation in protein biosynthesis is used by antibiotics such as: anisomycin, cycloheximide, chloramphenicol, tetracycline, streptomycin, erythromycin, puromycin etc.

EVENTS FOLLOWING PROTEIN TRANSLATION

The events following biosynthesis include post-translational modification and protein folding. During and after synthesis, polypeptide chains often fold to assume, so called, native secondary and tertiary structures. This is known as *protein folding*. Many proteins undergo *post-translational modification*. This may include the formation of disulfide bridges or attachment of any of a number of biochemical functional groups, such as acetate, phosphate, various lipids and carbohydrates. Enzymes may also remove one or more amino acids from the leading (amino) end of the polypeptide chain, leaving a protein consisting of two polypeptide chains connected by disulfide bonds.

PEPTIDE SYNTHESIS

In organic chemistry, peptide synthesis is the production of peptides, which are organic compounds in which multiple amino acids bind via peptide bonds which are also known as amide bonds. The biological process of producing long peptides (proteins) is known as protein biosynthesis.

Chemistry

Peptides are synthesized by coupling the carboxyl group or C-terminus of one amino acid to the amino group or N-terminus of another.

Liquid-phase Synthesis

Liquid-phase peptide synthesis is a classical approach to peptide synthesis. It has been replaced in most labs by solid-phase synthesis. However, it retains usefulness in large-scale production of peptides for industrial purposes.

Solid-phase Synthesis

Solid-phase peptide synthesis (SPPS), pioneered by Robert Bruce Merrifield, resulted in a paradigm shift within the peptide synthesis community. It is now the accepted method for creating peptides and proteins in the lab in a synthetic manner. SPPS allows the synthesis of natural peptides which are difficult to express in bacteria, the incorporation of unnatural amino acids, peptide/protein backbone modification, and the synthesis of D-proteins, which consist of D-amino acids.

Small solid beads, insoluble yet porous, are treated with functional units ('linkers') on which peptide chains can be built. The peptide will remain covalently attached to the bead until cleaved from it by a reagent such as trifluoroacetic acid. The peptide is thus 'immobilized' on the solid-phase and can be retained during a filtration process, whereas liquid-phase reagents and by-products of synthesis are flushed away.

The general principle of SPPS is one of repeated cycles of coupling-deprotection. The free N-terminal amine of a solid-phase attached peptide is coupled to a single N-protected amino acid unit. This unit is then deprotected, revealing a new N-terminal amine to which a further amino acid may be attached. The overwhelmingly important consideration is to generate extremely high yield in each step. For example, if each coupling step were to have 99% yield, a 26-amino acid peptide would be synthesized in 77% final yield (assuming 100% yield in each deprotection); if each step were 95%, it would be synthesized in 25% yield. Thus each amino acid is added in major excess (2~10x) and coupling amino acids together is highly optimized by a series of well-characterized agents.

There are two majorly used forms of SPPS – Fmoc and Boc. Unlike ribosome protein synthesis, solid-phase peptide synthesis proceeds in a C-terminal to N-terminal fashion. The N-termini of amino acid monomers is protected by these two groups and added onto a deprotected amino acid chain.

Automated synthesizers are available for both techniques, though many research groups continue to perform SPPS manually. SPPS is limited by yields, and typically peptides and proteins in the range of 70~100 amino acids are pushing the limits of synthetic accessibility. Synthetic difficulty also is sequence dependent; typically amyloid peptides and proteins are difficult to make. Longer lengths can be accessed by using native chemical ligation to couple two peptides together with quantitative yields.

t-Boc Solid-phase Peptide Synthesis

When Merrifield invented SPPS in 1963, it was according to the t-Boc method. t-Boc (or Boc) stands for (t)ert-(B)ut(o)xy(c)arbonyl. To remove Boc from a growing peptide chain, acidic conditions are used (usually neat TFA). Removal of side-chain protecting groups and the peptide from the resin at the end of the synthesis is achieved by incubating in hydrofluoric acid (which can be dangerous or even deadly); for this reason Boc chemistry is generally disfavored. Also, HF cleavage needs to be done in special fume hoods using specialized equipment. However for complex syntheses Boc is favourable. When synthesizing nonnatural peptide analogs which are base-sensitive (such as depsipeptides), Boc is necessary.

Fmoc Solid-phase Peptide Synthesis

This method was introduced by Carpino in 1972 and

further applied by Atherton in 1978. Fmoc stands for 9*H*-(f)luoren-9-yl(m)eth(o)xy(c)arbonyl which describes the Fmoc protecting group, first described as a protecting group by Carpino in 1970. To remove an Fmoc from a growing peptide chain, basic conditions (usually 20% piperidine in DMF) are used. Removal of side-chain protecting groups and peptide from the resin is achieved by incubating in trifluoroacetic acid (TFA), deionized water, and triisopropylsilane. Fmoc deprotection is usually slow because the anionic nitrogen produced at the end is not a particularly favorable product, although the whole process is thermodynamically driven by the evolution of carbon dioxide. The main advantage of Fmoc chemistry is that no hydrofluoric acid is needed. It is therefore used for most routine synthesis.

Bop Spps

The use of BOP reagent was first described by Castro et al in 1975.

SOLID SUPPORTS

The physical properties of the solid support, and the applications to which it can be utilized, vary with the material from which the support is constructed, the amount of crosslinking, as well as the linker and handle being used.

Polystyrene Resin

Polystyrene resin is a versatile resin and it is quite useful in multi-well, automated peptide synthesis, due to its minimal swelling in dichloromethane.

Polyamide Resin

Polyamide resin is also a useful and versatile resin. It seems to swell much more than polystyrene, in which case it may not be suitable for some automated synthesizers, if the wells are too small.

PEG Based Resin

ChemMatrix(R) is a new type of resin which is based on

PEG that is crosslinked. ChemMatrix(R) has claimed a high chemical and thermal stability (is compatible with Microwave synthesis) and has shown higher degrees of swellings in acetonitrile, dichloromethane, DMF, N-methyl pyrollidone, TFA and water compared to the polystyrene based resins. ChemMatrix has shown significant improvements to the synthesis of hydrophobic sequences. ChemMatrix is recommended for the synthesis of difficult and long peptides.

PROTECTING GROUPS

Due to amino acid excesses used to ensure complete coupling during each synthesis step, polymerization of amino acids is common in reactions where each amino acid is not protected. In order to prevent this polymerization, protecting groups are used. This adds additional deprotection phases to the synthesis reaction, creating a repeating design flow as follows:

- Protective group is removed from trailing amino acids in a deprotection reaction
- Deprotection reagents washed away to provide clean coupling environment
- Protected amino acids dissolved in a solvent such as dimethylformamide (DMF) are combined with coupling reagents are pumped through the synthesis column
- Coupling reagents washed away to provide clean deprotection environment

Currently, two protective groups (t-Boc, Fmoc) are commonly used in solid-phase peptide synthesis. Their lability is caused by the carbamate group which readily releases CO_2 for an irreversible decoupling step.

t-Boc Protective Group

The t-Boc group was commonly used for protecting the terminal amine of the peptide, requiring the use of more acid stable groups for side chain protection in orthogonal strategies. It retains usefulness in reducing aggregation of peptides during synthesis. Boc groups can be added to amino acids with t-Boc anhydride and a suitable base.

Fmoc Pprotective Group

Fmoc (9*H*-fluoren-9-ylmethoxycarbonyl) is currently a widely used protective group that is generally removed from the N terminus of a peptide in the iterative synthesis of a peptide from amino acid units. The advantage of Fmoc is that it is cleaved under very mild basic conditions (e.g. piperidine), but stable under acidic conditions. This allows mild acid labile protecting groups that are stable under basic conditions, such as Boc and benzyl groups, to be used on the side-chains of amino acid residues of the target peptide. This orthogonal protecting group strategy is common in the art of organic synthesis.

FMOC is preferred over BOC due to ease of cleavage; however it is less atom-economical, as the fluorenyl group is much larger than the tert-butyl group. Accordingly, prices for FMOC amino acids were high until the large-scale piloting of one of the first synthesized peptide drugs, enfuvirtide, began in the 1990s, when market demand adjusted the relative prices of the two sets of amino acids.

Benzyloxy-carbonyl (Z) Group

The first use of (Z) group as protecting groups was done by Max Bergmann who synthesised oligopeptides. Another carbamate based group is the benzyloxy-carbonyl (Z) group. It is removed in harsher conditions: HBr/acetic acid or catalytic hydrogenation. Today it is almost exclusively used for side chain protection.

Alloc Protecting Group

The allyloxycarbonyl (alloc) protecting group is often used to protect a carboxylic acid, hydroxyl, or amino group when an orthogonal deprotection scheme is required. It is sometimes used when conducting on-resin cyclic peptide formation, where the peptide is linked to the resin by a side-chain functional group. The alloc group can be removed using tetrakis (triphenylphosphine)palladium(0) along with a 37:2:1 mixture of chloroform, acetic acid, and N-methylmorpholine (NMM) for 2 hours. The resin must then be carefully washed 0.5% DIPEA in DMF, 3x10 ml of 0.5% sodium diethylthiocarbamate in DMF, and then 5x10 ml of 1:1 DCM:DMF.

Lithographic Protecting Groups

For special applications like protein microarrays lithographic protecting groups are used. Those groups can be removed through exposure to light.

ACTIVATING GROUPS

For coupling the peptides the carboxyl group is usually activated. This is important for speeding up the reaction. There are two main types of activating groups: carbodiimides and triazolols.

Carbodiimides

These activating agents were first developed. Most common are dicyclohexylcarbodiimide (DCC) and diisopropylcarbodiimide (DIC). Reaction with a carboxylic acid yields a highly reactive O-acyl-urea. During artificial protein synthesis (such as Fmoc solid-state synthesizers), the C-terminus is often used as the attachment site on which the amino acid monomers are added. To enhance the electrophilicity of carboxylate group, the negatively charged oxygen must first be "activated" into a better leaving group. DCC is used for this purpose.

The negatively charged oxygen will act as a nucleophile, attacking the central carbon in DCC. DCC is temporarily attached to the former carboxylate group (which is now an ester group), making nucleophilic attack by an amino group (on the attaching amino acid) to the former C-terminus (carbonyl group) more efficient. The problem with carbodiimides is that they are too reactive and that they can therefore cause racemization of the amino acid.

Triazolols

To solve the problem of racemization, triazolols were introduced. The most important ones are 1-hydroxy-benzotriazole (HOBt) and 1-hydroxy-7-aza-benzotriazole

(HOAt). Others have been developed. These substances can react with the O-acylurea to form an active ester which is less reactive and less in danger of racemization. HOAt is especially favourable because of a neighbouring group effect.

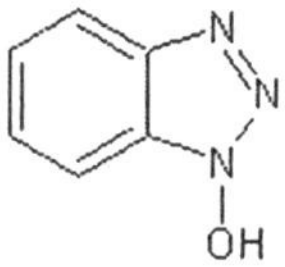

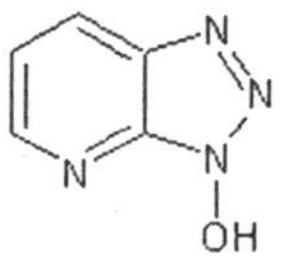

N N N OH N N N N OH

Newer developments omit the carbodiimides totally. The active ester is introduced as a uronium or phosphonium salt of a non-nucleophilic anion (tetrafluoroborate or hexafluorophosphate): HBTU, HATU, PyBOP.

SYNTHESIZING LONG PEPTIDES

Stepwise elongation, in which the amino acids are connected step-by-step in turn, is ideal for small peptides containing between 2 and 100 amino acid residues. Another method is fragment condensation, in which peptide fragments are coupled. Although the former can elongate the peptide chain without racemization, the yield drops if only it is used in the creation of long or highly polar peptides. Fragment condensation is better than stepwise elongation for synthesizing sophisticated long peptides, but its use must be restricted in order to protect against racemization. Fragment condensation is also undesirable since the coupled fragment must be in gross excess, which may be a limitation depending on the length of the fragment.

A new development for producing longer peptide chains is chemical ligation: Unprotected peptide chains react chemoselectively in aqueous solution. A first kinetically controlled product rearranges to form the amide bond. The most common form of native chemical ligation uses a peptide thioester that reacts with a terminal cystein residue.

MICROWAVE ASSISTED PEPTIDE SYNTHESIS

Although microwave irradiation has been around since the late 1940s, it was not until 1986 that microwave energy was used in organic chemistry. During the end of the 1980s and 1990s, microwave energy was an obvious source for completing chemical reactions in minutes that would otherwise take several hours to days. Through several technical improvements at the end of the 1990s and beginning of the 2000s, microwave synthesizers have been designed to provide both low and high energy pockets of microwave energy so that the temperature of the reaction mixture could controlled. The microwave energy used in peptide synthesis is of a single frequency providing maximum penetration depth of the sample which is in contrast to conventional kitchen microwaves. In peptide synthesis, microwave irradiation has been used to complete long peptide sequences with high degrees of yield and low degrees of racemization. Microwave irradiation during the coupling of amino acids to a growing polypeptide chain is not only catalyzed through the increase in temperature, but also due to the alternating electromagnetic radiation to which the polar backbone of the polypeptide continuously aligns to.

Due to the this phenomenon, the microwave energy can prevent aggregation and thus increases yields of the final peptide product. Despite the main advantages of microwave irradiation of peptide synthesis, the main disadvantage is the racemization which may occur with the coupling of cysteine and histidine. A typical coupling reaction with these amino acids are performed at lower temperatures than the other 18 natural amino acids. Another disadvantage is that allyl containing amino acid derivatives cannot be coupled to amino acids using microwave irradiation due to uncontrolled polymerization.

CYCLIC PEPTIDES

AN EXAMPLE OF SOLID PHASE PEPTIDE SYNTHESIS

The following is an outline of the synthetic steps for peptide synthesis on polyamide or polystyrene resin, using the base labile 9*H*-fluoren-9-ylmethoxycarbonyl (Fmoc) protecting group. Using the techniques outlined below, one will obtain a peptide which is capped on the N-terminus with and acetyl group, and on the C-terminus with a primary amide ($CONH_2$).

Setting Up Glassware for Manual Peptide Synthesis

Manual peptide synthesis can be accomplished in a fritted-filter reaction vessel with a three-way valve fitted onto a 1 L side arm vacuum flask by way of a 1-hole stopper. One valve is used to bubble nitrogen, which is first passed through a small column of Drierite, and then into the reaction mixture to agitate the solution and mix reagents. The other valve is used to evacuate excess reaction solutions and wash solvent using a vacuum flask. All glass pieces to be used in Solid-phase synthesis should be treated with a silanizing agent (such as 1-5% dimethyldichlorosilane in DCM) prior to use, to avoid accumulation of static charge, which makes the resin very difficult to handle.

Preparation of Polyamide-Rink Resin

Polyamide (PL-DMA) resin (1g) is treated with ethylene diamine (40 ml) in a 50 ml Falcon tube overnight on a rocker, then filtered, washed with 5x10 ml of 1:1 dimethylformamide (DMF):dichloromethane (DCM) solution, 5x10 ml of 1:1 DCM, and loaded with Fmoc-Rink using Benzotriazol-1-yl-oxytripyrrolidinophosphonium hexafluorophosphate (PyBOP) (3 eq), 1-Hydroxybenzotriazole Hydrate (HOBt) (3 eq), and Diisopropylethylamine (DIPEA) (6 eq) in 1:1 DCM:DMF. It can then be dried under vacuum and stored at "15 °C until needed.

Handling the Resin before, and during Synthesis

The resin is first swelled for 15 minutes in 10 ml of 1:1 DCM:DMF and drained. The resin is also washed with 5x10 ml of 1:1 DCM and DMF after each completed amino acid coupling.

Fmoc-Deprotection

Fmoc deprotection after each amino acid coupling is accomplished using 2x10 ml of 20% piperidine in DMF, with N2 agitation for 10 minutes each treatment. The resin is then washed with 5x10 ml DMF, followed by 5x10 ml of 1:1

DCM:DMF. An alternate treatment is 1% DBU in DMF; this gentler treatment allows removal of Fmoc groups in the presence of other base-labile moieties.

Adding Amino Acids

An amino acid is coupled to the deprotected N-terminal amine of the resin, or previously coupled amino acid, using a coupling mixture such as the protected amino acid (3 eq), PyBOP (3 eq), HOBt (3 eq), and DIPEA (6 eq) in 1:1 DCM:DMF until the resin is negative to ninhydrin. Another popular set of coupling conditions is amino acid (4.4 eq), HBTU (4 eq), and DIPEA (8 eq) in DMF. Amino acids can also be purchased as the pre-activated ester, in which case coupling agents such as HBTU and PyBOP are unnecessary.

Monitoring the Progress of Amino Acid Couplings

The progress of amino acid couplings can be followed using ninhydrin, or p-chloranil. The ninhydrin solution turns dark blue (positive result) in the presence of a free primary amine but is otherwise colorless (negative result). The p-chloranil solution will turn the resin beads dark black or blue in the presence of a primary amine if acetaldehyde is used as the solvent or in the presence of a secondary amine, if acetone is used instead; the beads remain colorless or pale yellow otherwise.

Testing by Ninhydrin (1)

Add 2 drops of 40% phenol in ethanol, 2 drops of 0.014 mol/L KCN in pyridine, and 4 drops of 5% ninhydrin in ethanol to a microcentrifuge tube along with a spatula tip size sample of resin, then vortex the mixture and heat for 5 minutes at 100 °C.

Testing by Chloranil (2)

Add 5 drops of acetone or acetaldehyde, 5 drops of a saturated solution of p-chloranil in toluene, plus a small spatula-tip-size sample of resin to a microcentrifuge tube, then vortex the mixture and allow to stand at room temperature for 5 minutes. Acetone is used for the detection of secondary amines, where acetaldehyde is used for primary amines.

Continuing Peptide Extension: Once the coupling of the amino acid is complete, the resin is washed, the Fmoc group deprotected with piperidine, and the resin washed again to prepare it for the next coupling. This process is repeated until all necessary amino acids have been added.

Acetylating the N-terminus: After the peptide sequence is completed, the N-terminal amine can be acetylated with 2 ml of 1:1 acetic anhydride and triethylamine in 10 ml of 1:1 DCM:DMF for 1 hour or until negative to ninhydrin, and the resin then washed with 5x10 ml of 1:1 DCM:DMF, before the peptide is cleaved from the resin. The N-terminus can also be left as the free amine if required.

Cleaving the Peptide from the Resin

The resin is treated with a cocktail of trifluoroacetic acid (TFA) and cleavage scavenger reagents, such as triisopropylsilane (TIPS), 1,2-ethanedithiol, and water (H2O). The choice of scavengers is dependent on the amino acid sequence. The resin is then filtered away, and the combined filtrates allowed to stand for 1 hour to ensure removal of the acid labile protecting groups.

Workup of Peptides After Cleavage from the Resin

The TFA is evaporated to dryness (or a heavy oil or glass if it does not solidify) on the rotary evaporator, followed by the addition of 5 ml of diethyl ether to the flask to precipitate the peptide, and remove the bulk of the by-products. Once the peptide is precipitated with the ether, filter through a sintered glass funnel and redissolve peptide into 60% AcN/ 0.1 %TFA. The peptide solution is then frozen in a dry ice/ ethanol bath and lyophilised (dried). The result is a crude peptide which is stable for a number of years. The peptide can then be purified by Ion-exhange and Reverse Phase chromatography. Typically preparative HPLC is used to purify the final product. Mass spectrometry data is obtained to ensure the target peptide was obtained.

Chapter 10

Biophysics

Biophysics is the application of physics to biological systems. This interdisciplinary field is quickly growing in the modern scientific community. The Department of Biophysics, is a centre of drug discovery and clinical proteomics. It has combined the fields of structural biology, bioinformatics and proteomics seamlessly. The goal of modern research in Drug Discovery is to develop drugs that will act in a specific way with minimal side effects and also are demonstrably better than the existing therapies.

The conventional approaches of Drug discovery render it a long and an expensive process. Biophysics is an interdisciplinary field which applies techniques from the physical sciences to understanding biological structure and function at the molecular level. In simpler terms we use a range of scientific techniques to try to understand what macromolecules like DNA, proteins, fats and sugars look like, and how they interact to form the biological systems around us.

Plants, animals and even seemingly simple organisms like bacteria and archae consist of complex biological networks of chemical signals and molecular interactions, which enable them to do such things as move, respire or reproduce. Biophysics is a varied field which draws on biology, physics, chemistry, mathematics, engineering, genetics, physiology and medicine, all with the aim to understand these systems using experiments or theoretical and computational modeling. It is also a young science and is still rapidly developing - bionanotechnology and biosensors are just some of the latest fields to emerge in the last few years. It is also known as biological physics that applies the theories and methods of the physical sciences to questions of biology.

Biophysics apply to theories of the human species and methods of the way we question problems. Biophysics research today is comprised of several specific biological studies which neither share a unique identifying factor nor subject themselves to clear and concise definitions. The studies included under the umbrella of biophysics range from sequence analysis to neural networks. Biophysics is also concerned with creating mechanical limbs and nanomachines to regulate biological functions, although currently these are more commonly referred to as belonging to the fields of bioengineering and nanotechnology respectively. Biophysics typically addresses biological questions that are similar to those in biochemistry, but the questions are asked at a molecular level.

Traditional studies in biochemistry and molecular biology are conducted using statistical ensemble experiments, typically using pico- to micro molar concentrations of macromolecules. Because the molecules that comprise living cells are so small, techniques such as PCR amplification, gel blotting, fluorescence labeling and in vivo staining are used so that experimental results are observable with an unaided eye or, at most, optical magnification.

Using these techniques, researchers in these subjects attempt to elucidate the complex systems of interactions that give rise to the processes that make life possible. By drawing knowledge and experimental techniques from a wide variety of disciplines, biophysicists are able to indirectly observe or model the structures and interactions of *individual* molecules or complexes of molecules. They require screening of hundreds of thousands of samples before reaching some potential compounds with desired properties. By some estimates, it takes dozens of years and millions of dollars.

However, with the advances in protein structure determination, structure based drug design has emerged as a powerful tool for developing new drugs with specific properties and minimal side effects. This technique is usually faster than the conventional methods. In structure based drug design, the three-dimensional structure of a drug target interacting with small molecules is used for drug discovery. Structure-based drug design represents the idea that one can see exactly how the ligand molecule interacts with its target protein. The designed compounds that have affinities in the acceptable pharmacological range can be further processed for other biological assays and clinical trials.

BIOPHYSICS: NEW CHALLENGES FOR PHYSICS

Recent progress in life sciences has demonstrated that the first decades of the new century are likely to be dominated by developments in this field. Since physics forms the basis of the life science evolution, the visionary guidance and assistance of physicists who have enjoyed a versatile training will be needed. Perhaps the most easily recognizable example of this need for an interdisciplinary approach is the astounding revelation of the human genetic code, the self-organized plan according to which self-reproducing entities, such as cells, form complex organisms.

The underlying interactions are governed by physical principles. An equally important challenge to physics is to clarify the way in which ensembles of cells, cells as individual entities, and their molecular constituents, function in their respective surroundings. Biophysics has thus become a central theme of the physics department, offering young students unique opportunities for study and ample openings for top-level research.

Excellent job opportunities and the dynamics of Munich as a centre of biotechnology with its many startup companies underline the attractiveness of the biophysics programme of the TU Munich (TUM). The role of physical methods in life sciences is manifested by modern techniques such as ultrasound, positron emission, X-ray and nuclear magnetic resonance tomography, light and electron microscopy, laser spectroscopy, X-ray structure analysis, and electrophysiological techniques such as patch clamp.

These techniques have benefited from more than half a

century of research in physics and are the result of combining classical instrumentation with computational physics. The discoveries of these new physical methods have triggered off dramatic progress in life sciences. At the same time there are also numerous examples originating from biology that have inspired new developments in physics.

The most prominent one is the discovery of the general energy conservation law by Robert Meyer and Hermann von Helmholtz and the theory of Brownian motion by Albert Einstein. Einstein's ingenious interpretation of the observation of the botanist Robert Brown that seeds perform random walks in water influenced the development of modern physics at the beginning of the last century nearly as much as Planck's equation describing black body radiation. Nowadays, physicists do not content themselves with being the designers of new instrumentation but strive for a more active role in the search for universal physical principles governing the assembly and function of biomaterials.

In order to be successful, it is absolutely necessary that physicists accept the complexity of biomaterials and become familiar with the principal questions of biology. The physicist's capacity to unravel universal laws governing complex processes or to develop new measuring techniques to test laws predicted by theory is urgently needed in life sciences.

BIOPHYSICS AT THE TUM

As early as 1980, to meet the challenges of modern life sciences, the physics department of the TUM decided to strengthen its resources in the field of biophysics. There are now 5 full professors at the TUM, complemented by 3 adjunct professors, all working in the field of biophysics. Recently, a junior professorship and a junior group funded by the Emmy Noether programme were established so as to broaden the scope of the research activities.

The TUM was, in fact, the first university in Germany to introduce a full biophysics programme. The combination of experimental and theoretical studies on single molecules, living cells, and cellular networks is a unique strength of the biophysical research efforts of the physics department. The goal is not only to obtain a detailed understanding of biophysical principles but also to develop new tools with which to study and quantify biological processes. It is this combination of different approaches and foci that makes the study of biophysics at the TUM an outstanding experience. The biophysics research groups of the physics department engage in a very broad spectrum of activities and are complemented by several special areas of research.

BIOCHEMICAL TECHNIQUE

FUSION PROTEIN

Fusion proteins, also known as chimeric proteins, are proteins created through the joining of two or more genes which originally coded for separate proteins. Translation of this *fusion gene* results in a single polypeptide with function properties derived from each of the original proteins. *Recombinant.fusion proteins* are created artificially by recombinant DNA technology for use in biological research or therapeutics.

Chimeric mutant proteins occur naturally when a large-scale mutation, typically a chromosomal translocation, creates a novel coding sequence containing parts of the coding sequences from two different genes. Naturally occurring fusion proteins are important in cancer, where they may function as oncoproteins. The bcr-abl fusion protein is a well-known example of an oncogenic fusion protein, and is considered to be the primary oncogenic driver of chronic myelogenous leukemia.

Properties of Fusion Proteins

The functionality of fusion proteins is made possible by the fact that many protein functional domains are *modular*. In other words, the linear portion of a polypeptide which corresponds to a given domain, such as a tyrosine kinase domain, may be removed from the rest of the protein without destroying its intrinsic enzymatic capability.

Recombinant fusion proteins

A recombinant fusion protein is a protein created through genetic engineering of a fusion gene. This typically involves removing the stop codon from a cDNA sequence coding for the first protein, then appending the cDNA sequence of the second protein in frame through ligation or overlap extension PCR. That DNA sequence will then be expressed by a cell as a single protein.

The protein can be engineered to include the full sequence of both original proteins, or only a portion of either. If the two entities are proteins, often linker (or "spacer") peptides are also added which make it more likely that the proteins fold independently and behave as expected.

Especially in the case where the linkers enable protein purification, linkers in protein or peptide fusions are sometimes engineered with cleavage sites for proteases or chemical agents which enable the liberation of the two separate proteins.

This technique is often used for identification and purification of proteins, by fusing a GST protein, FLAG peptide, or a hexa-his peptide (aka: a 6xhis-tag) which can be isolated using nickel or cobalt resins (affinity chromatography). Chimeric proteins can also be manufactured with toxins or anti-bodies attached to them in order to study disease development.

Chimeric Protein Drugs

Several drugs made from chimeric proteins are currently available for medical use. Several chimeric protein drugs are TNFá blockers, such as Etanercept, Infliximab, and Adalimumab.

NATURALLY OCCURRING FUSION PROTEINS

Naturally occurring fusion genes are most commonly created when a chromosomal translocation replaces the terminal exons of one gene with intact exons from a second gene. This creates a single gene which can be transcribed, spliced, and translated to produce a functional fusion protein.

Many important cancer-promoting oncogenes are fusion genes produced in this way.

Examples include:

- Gag-onc fusion protein
- Bcr-abl fusion protein
- Tpr-met fusion protein

POLYMERASE CHAIN REACTION

The polymerase chain reaction (PCR) is a technique widely used in molecular biology. It derives its name from one of its key components, a DNA polymerase used to amplify a piece of DNA by *in vitro* enzymatic replication. As PCR progresses, the DNA thus generated is itself used as a template for replication. This sets in motion a chain reaction in which the DNA template is exponentially amplified.

With PCR it is possible to amplify a single or few copies of a piece of DNA across several orders of magnitude, generating millions or more copies of the DNA piece. PCR can be extensively modified to perform a wide array of genetic manipulations. Almost all PCR applications employ a heat-stable DNA polymerase, such as Taq polymerase, an enzyme originally isolated from the bacterium *Thermus aquaticus*. This DNA polymerase enzymatically assembles a new DNA strand from DNA building blocks, the nucleotides, using single-stranded DNA as template and DNA oligonucleotides (also called DNA primers) required for initiation of DNA synthesis.

The vast majority of PCR methods use thermal cycling, i.e., alternately heating and cooling the PCR sample to a defined series of temperature steps. These thermal cycling steps are necessary to physically separate the strands (at high temperatures) in a DNA double helix (DNA melting) used as template during DNA synthesis (at lower temperatures) by the DNA polymerase to selectively amplify the target DNA.

The selectivity of PCR results from the use of primers that are complementary to the DNA region targeted for amplification under specific thermal cycling conditions. Developed in 1983 by Kary Mullis, PCR is now a common and often indispensable technique used in medical and biological research labs for a variety of applications. These include DNA cloning for sequencing, DNA-based phylogeny, or functional analysis of genes; the diagnosis of hereditary diseases; the identification of genetic fingerprints (used in forensic sciences and paternity testing); and the detection and diagnosis of infectious diseases. In 1993 Mullis won the Nobel Prize in Chemistry for his work on PCR.

PCR PRINCIPLES AND PROCEDURE

PCR is used to amplify specific regions of a DNA strand (the DNA target). This can be a single gene, a part of a gene, or a non-coding sequence. Most PCR methods typically amplify DNA fragments of up to 10 kilo base pairs (kb), although some techniques allow for amplification of fragments up to 40 kb in size. A basic PCR set up requires several components and reagents. These components include:

- *DNA template* that contains the DNA region (target) to be amplified.
- Two *primers*, which are complementary to the DNA regions at the 5' (five prime) or 3' (three prime) ends of the DNA region.
- A DNA polymerase such as *Taq polymerase* or another DNA polymerase with a temperature optimum at around 70°C.
- *Deoxynucleoside triphosphates* (dNTPs; also very commonly and erroneously called deoxynucleotide triphosphates), the building blocks from which the DNA polymerases synthesizes a new DNA strand.
- *Buffer solution,* providing a suitable chemical environment for optimum activity and stability of the DNA polymerase.
- *Divalent cations,* magnesium or manganese ions; generally Mg^{2+} is used, but Mn^{2+} can be utilized for PCR-mediated DNA mutagenesis, as higher Mn^{2+} concentration increases the error rate during DNA synthesis
- *Monovalent cation* potassium ions.

The PCR is commonly carried out in a reaction volume of 10-200 ìl in small reaction tubes (0.2-0.5 ml volumes) in a thermal cycler. The thermal cycler heats and cools the reaction tubes to achieve the temperatures required at each step of the reaction. Many modern thermal cyclers make use of the Peltier effect which permits both heating and cooling of the block holding the PCR tubes simply by reversing the electric current.

Thin-walled reaction tubes permit favorable thermal conductivity to allow for rapid thermal equilibration. Most thermal cyclers have heated lids to prevent condensation at the top of the reaction tube. Older thermocyclers lacking a heated lid require a layer of oil on top of the reaction mixture or a ball of wax inside the tube.

Procedure

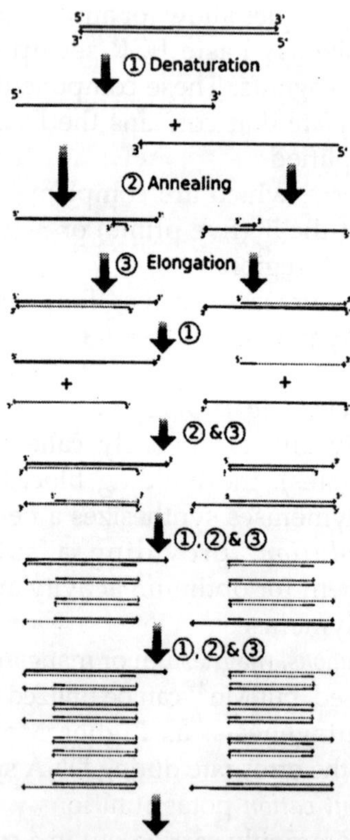

Fig. Schematic Drawing of the PCR Cycle

Denaturing at 94-96°C. (2) Annealing at~65°C (3) Elongation at 72°C. Four cycles are shown here. The blue lines represent the DNA template to which primers (red arrows) anneal that are extended by the DNA polymerase (light green circles), to give shorter DNA products (green lines), which themselves are used as templates as PCR progresses.

The PCR usually consists of a series of 20 to 40 repeated temperature changes called cycles; each cycle typically consists of 2-3 discrete temperature steps. Most commonly PCR is carried out with cycles that have three temperature steps. The cycling is often preceded by a single temperature step (called *hold*) at a high temperature (>90°C), and followed by one hold at the end for final product extension or brief storage. The temperatures used and the length of time they are applied in each cycle depend on a variety of parameters.

These include the enzyme used for DNA synthesis, the concentration of divalent ions and dNTPs in the reaction, and the melting temperature (Tm) of the primers.

Initialization step

This step consists of heating the reaction to a temperature of 94-96°C (or 98°C if extremely thermostable polymerases are used), which is held for 1-9 minutes. It is only required for DNA polymerases that require heat activation by hot-start PCR.

Denaturation step

This step is the first regular cycling event and consists of heating the reaction to 94-98°C for 20-30 seconds. It causes melting of DNA template and primers by disrupting the hydrogen bonds between complementary bases of the DNA strands, yielding single strands of DNA.

Annealing step

The reaction temperature is lowered to 50-65°C for 20-40 seconds allowing annealing of the primers to the single-stranded DNA template. Typically the annealing temperature is about 3-5 degrees Celsius below the Tm of the primers used. Stable DNA-DNA hydrogen bonds are only formed when the primer sequence very closely matches the template sequence. The polymerase binds to the primer-template hybrid and begins DNA synthesis.

Extension/elongation step

The temperature at this step depends on the DNA polymerase used; Taq polymerase has its optimum activity temperature at 75-80°C, and commonly a temperature of 72°C is used with this enzyme. At this step the DNA polymerase synthesizes a new DNA strand complementary to the DNA template strand by adding dNTPs that are complementary to the template in 5' to 3' direction, condensing the 5'-phosphate group of the dNTPs with the 3'-hydroxyl group at the end of the nascent (extending) DNA strand.

The extension time depends both on the DNA polymerase used and on the length of the DNA fragment to be amplified. As a rule-of-thumb, at its optimum temperature, the DNA polymerase will polymerize a thousand bases per minute. Under optimum conditions, i.e., if there are no limitations due to limiting substrates or reagents, at each extension step, the amount of DNA target is doubled, leading to exponential (geometric) amplification of the specific DNA fragment.

Final elongation

This single step is occasionally performed at a temperature of 70-74°C for 5-15 minutes after the last PCR cycle to ensure that any remaining single-stranded DNA is fully extended.

Final hold

This step at 4-15°C for an indefinite time may be employed for short-term storage of the reaction. To check whether the PCR generated the anticipated DNA fragment (also sometimes referred to as the amplimer or amplicon), agarose gel electrophoresis is employed for size separation of the PCR products. The size(s) of PCR products is determined by comparison with a *DNA ladder* (a molecular weight marker), which contains DNA fragments of known size, run on the gel alongside the PCR products.

PCR STAGES

The PCR process can be divided into three stages:

Exponential amplification: At every cycle, the amount of product is doubled (assuming 100% reaction efficiency). The reaction is very specific and precise.

Levelling off stage: The reaction slows as the DNA polymerase loses activity and as consumption of reagents such as dNTPs and primers causes them to become limiting.

Plateau: No more product accumulates due to exhaustion of reagents and enzyme.

PCR Optimization

In practice, PCR can fail for various reasons, in part due to its sensitivity to contamination causing amplification of spurious DNA products. Because of this, a number of techniques and procedures have been developed for optimizing PCR conditions. Contamination with extraneous DNA is addressed with lab protocols and procedures that separate pre-PCR mixtures from potential DNA contaminants.

This usually involves spatial separation of PCR-setup areas from areas for analysis or purification of PCR products, and thoroughly cleaning the work surface between reaction setups. Primer-design techniques are important in improving PCR product yield and in avoiding the formation of spurious products, and the usage of alternate buffer components or polymerase enzymes can help with amplification of long or otherwise problematic regions of DNA.

APPLICATION OF PCR

ISOLATION OF GENOMIC DNA

PCR allows isolation of DNA fragments from genomic DNA by selective amplification of a specific region of DNA. This use of PCR augments many methods, such as generating hybridization probes for Southern or northern hybridization and DNA cloning, which require larger amounts of DNA, representing a specific DNA region.

PCR supplies these techniques with high amounts of pure

DNA, enabling analysis of DNA samples even from very small amounts of starting material. Other applications of PCR include DNA sequencing to determine unknown PCR-amplified sequences in which one of the amplification primers may be used in Sanger sequencing, isolation of a DNA sequence to expedite recombinant DNA technologies involving the insertion of a DNA sequence into a plasmid or the genetic material of another organism.

Bacterial colonies (E.coli) can be rapidly screened by PCR for correct DNA vector constructs. PCR may also be used for genetic fingerprinting; a forensic technique used to identify a person or organism by comparing experimental DNAs through different PCR-based methods.

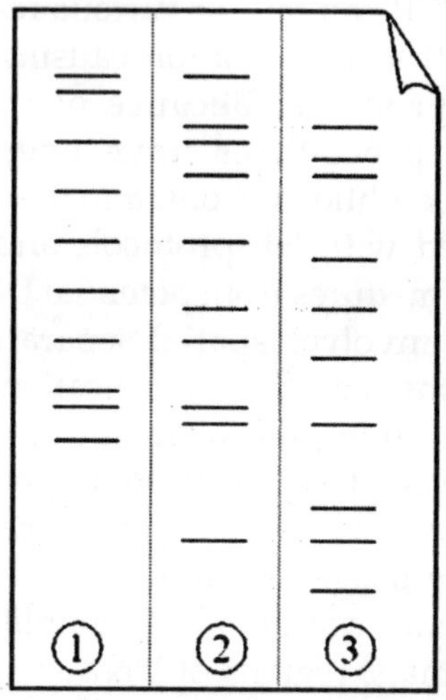

Fig. Electrophoresis of PCR-amplified DNA Fragments. (1) Father. (2) Child. (3) Mother. The Child has Inherited some, but not all of the Fingerprint of each of its Parents, giving it a new, Unique Fingerprint

Some PCR 'fingerprints' methods have high discriminative power and can be used to identify genetic relationships between individuals, such as parent-child or between siblings, and are used in paternity testing. This technique may also be used to determine evolutionary relationships among organisms.

AMPLIFICATION AND QUANTITATION OF DNA

Because PCR amplifies the regions of DNA that it targets, PCR can be used to analyse extremely small amounts of sample. This is often critical for forensic analysis, when only a trace amount of DNA is available as evidence. PCR may also be used in the analysis of ancient DNA that is thousands of years old. These PCR-based techniques have been successfully used on animals, such as a forty-thousand-year-old mammoth, and also on human DNA, in applications ranging from the analysis of Egyptian mummies to the identification of a Russian Tsar.

Quantitative PCR methods allow the estimation of the amount of a given sequence present in a sample – a technique often applied to quantitatively determine levels of gene expression.

Real-time PCR is an established tool for DNA quantification that measures the accumulation of DNA product after each round of PCR amplification.

PCR IN DIAGNOSIS OF DISEASES

PCR allows early diagnosis of malignant diseases such as leukemia and lymphomas, which is currently the highest developed in cancer research and is already being used routinely. PCR assays can be performed directly on genomic DNA samples to detect translocation-specific malignant cells at a sensitivity which is at least 10,000 fold higher than other methods. PCR also permits identification of non-cultivatable or slow-growing microorganisms such as mycobacteria, anaerobic bacteria, or viruses from tissue culture assays and animal models.

The basis for PCR diagnostic applications in microbiology is the detection of infectious agents and the discrimination of non-pathogenic from pathogenic strains by virtue of specific genes.

Viral DNA can likewise be detected by PCR. The primers used need to be specific to the targeted sequences in the DNA of a virus, and the PCR can be used for diagnostic analyses or DNA sequencing of the viral genome. The high sensitivity of

PCR permits virus detection soon after infection and even before the onset of disease. Such early detection may give physicians a significant lead in treatment. The amount of virus ("viral load") in a patient can also be quantified by PCR-based DNA quantitation techniques.

GEL ELECTROPHORESIS

Gel electrophoresis is a technique used for the separation of deoxyribonucleic acid, ribonucleic acid, or protein molecules using an electric current applied to a gel matrix. It is usually performed for analytical purposes, but may be used as a preparative technique prior to use of other methods such as mass spectrometry, RFLP, PCR, cloning, DNA sequencing, or Southern blotting for further characterization.

SEPARATION

The term "gel" in this instance refers to the matrix used to contain, then separate the target molecules. In most cases the gel is a crosslinked polymer whose composition and porosity is chosen based on the specific weight and composition of the target to be analyzed.

When separating proteins or small nucleic acids (DNA, RNA, or oligonucleotides) the gel is usually composed of different concentrations of acrylamide and a cross-linker, producing different sized mesh networks of polyacrylamide.

When separating larger nucleic acids (greater than a few hundred bases), the preferred matrix is purified agarose. In both cases, the gel forms a solid, yet porous matrix. Acrylamide, in contrast to polyacrylamide, is a neurotoxin and must be handled using appropriate safety precautions to avoid poisoning.

"Electrophoresis" refers to the electromotive force (EMF) that is used to move the molecules through the gel matrix. By placing the molecules in wells in the gel and applying an electric current, the molecules will move through the matrix at different rates, usually determined by mass, toward the positive anode if negatively charged or toward the negative cathode if positively charged.

VISUALIZATION

After the electrophoresis is complete, the molecules in the gel can be stained to make them visible. Ethidium bromide, silver, or coomassie blue dye may be used for this process. Other methods may also be used to visualize the separation of the mixture's components on the gel. If the analyte molecules fluoresce under ultraviolet light, a photograph can be taken of the gel under ultraviolet lighting conditions. If the molecules to be separated contain radioactivity added for visibility, an autoradiogram can be recorded of the gel.

If several mixtures have initially been injected next to each other, they will run parallel in individual lanes. Depending on the number of different molecules, each lane shows separation of the components from the original mixture as one or more distinct bands, one band per component. Incomplete separation of the components can lead to overlapping bands, or to indistinguishable smears representing multiple unresolved components.

Bands in different lanes that end up at the same distance from the top contain molecules that passed through the gel with the same speed, which usually means they are approximately the same size. There are molecular weight size markers available that contain a mixture of molecules of known sizes. If such a marker was run on one lane in the gel parallel to the unknown samples, the bands observed can be compared to those of the unknown in order to determine their size. The distance a band travels is approximately inversely proportional to the logarithm of the size of the molecule.

APPLICATIONS

Gel electrophoresis is used in forensics, molecular biology, genetics, microbiology and biochemistry. The results can be analyzed quantitatively by visualizing the gel with UV light and a gel imaging device. The image is recorded with a computer operated camera, and the intensity of the band or spot of interest is measured and compared against standard or markers loaded on the same gel. The measurement and analysis are mostly done with specialized software.

Depending on the type of analysis being performed, other techniques are often implemented in conjunction with the results of gel electrophoresis, providing a wide range of field-specific applications.

Nucleic Acids

In the case of nucleic acids, the direction of migration, from negative to positive electrodes, is due to the naturally-occurring negative charge carried by their sugar-phosphate backbone.

Double-stranded DNA fragments naturally behave as long rods, so their migration through the gel is relative to their radius of gyration, or, for non-cyclic fragments, size. Single-stranded DNA or RNA tend to fold up into molecules with complex shapes and migrate through the gel in a complicated manner based on their tertiary structure. Therefore, agents that disrupt the hydrogen bonds, such as sodium hydroxide or formamide, are used to denature the nucleic acids and cause them to behave as long rods again.

Gel electrophoresis of large DNA or RNA is usually done by agarose gel electrophoresis.

Proteins

Proteins, unlike nucleic acids, can have varying charges and complex shapes, therefore they may not migrate into the gel at similar rates, or at all, when placing a negative to positive EMF on the sample. Proteins therefore, are usually denatured in the presence of a detergent such as sodium dodecyl sulfate/ sodium dodecyl phosphate (SDS/SDP) that coats the proteins with a negative charge. Generally, the amount of SDS bound is relative to the size of the protein (usually 1.4g SDS per gram of protein), so that the resulting denatured proteins have an overall negative charge, and all the proteins have a similar charge to mass ratio.

Since denatured proteins act like long rods instead of having a complex tertiary shape, the rate at which the resulting SDS coated proteins migrate in the gel is relative only to its size and not its charge or shape. Proteins are usually analyzed by sodium dodecyl sulfate polyacrylamide gel electrophoresis (SDS-PAGE), by native gel electrophoresis, by quantitative preparative native continuous polyacrylamide gel electrophoresis (QPNC-PAGE), or by 2-D electrophoresis.

DNA SEQUENCING

The term DNA sequencing encompasses biochemical methods for determining the order of the nucleotide bases, adenine, guanine, cytosine, and thymine, in a DNA oligonucleotide. The sequence of DNA constitutes the heritable genetic information in nuclei, plasmids, mitochondria, and chloroplasts that forms the basis for the developmental programs of all living organisms.

Determining the DNA sequence is therefore useful in basic research studying fundamental biological processes, as well as in applied fields such as diagnostic or forensic research. The advent of DNA sequencing has significantly accelerated biological research and discovery. The rapid speed of sequencing attained with modern DNA sequencing technology has been instrumental in the large-scale sequencing of the human genome, in the Human Genome Project.

Related projects, often by scientific collaboration across continents, have generated the complete DNA sequences of many animal, plant, and microbial genomes.

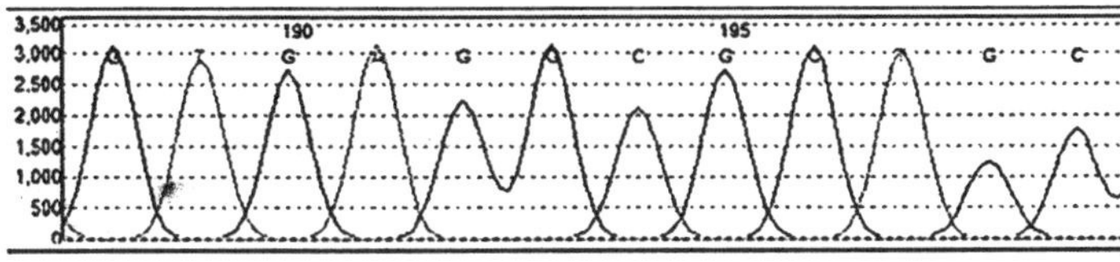

Fig. DNA Sequence Trace

Maxam-Gilbert Sequencing

In 1976-1977, Allan Maxam and Walter Gilbert developed a DNA sequencing method based on chemical modification of DNA and subsequent cleavage at specific bases. Although Maxam and Gilbert published their chemical sequencing method two years after the ground-breaking paper of Sanger and Coulson on plus-minus sequencing, Maxam-Gilbert sequencing rapidly became more popular, since purified DNA could be used directly, while the initial Sanger method required that each read start be cloned for production of single-stranded DNA.

However, with the development and improvement of the chain-termination method, Maxam-Gilbert sequencing has fallen out of favour due to its technical complexity, extensive use of hazardous chemicals, and difficulties with scale-up. In addition, unlike the chain-termination method, chemicals used in the Maxam-Gilbert method cannot easily be customized for use in a standard molecular biology kit.

In brief, the method requires radioactive labelling at one end and purification of the DNA fragment to be sequenced. Chemical treatment generates breaks at a small proportion of one or two of the four nucleotide bases in each of four reactions (G, A+G, C, C+T). Thus a series of labelled fragments is generated, from the radiolabelled end to the first 'cut' site in each molecule.

The fragments are then size-separated by gel electrophoresis, with the four reactions arranged side by side. To visualize the fragments generated in each reaction, the gel is exposed to X-ray film for autoradiography, yielding an image of a series of dark 'bands' corresponding to the radiolabelled DNA fragments, from which the sequence may be inferred.

Also sometimes known as 'chemical sequencing', this method originated in the study of DNA-protein interactions (footprinting), nucleic acid structure and epigenetic modifications to DNA, and within these it still has important applications.

Chain-termination Methods

While the chemical sequencing method of Maxam and Gilbert, and the plus-minus method of Sanger and Coulson were orders of magnitude faster than previous methods, the chain-terminator method developed by Sanger was even more efficient, and rapidly became the method of choice. The Maxam-Gilbert technique requires the use of highly toxic chemicals, and large amounts of radiolabeled DNA, whereas the chain-terminator method uses fewer toxic chemicals and lower amounts of radioactivity.

The key principle of the Sanger method was the use of dideoxynucleotides triphosphates (ddNTPs) as DNA chain terminators. The classical chain-termination or Sanger method requires a single-stranded DNA template, a DNA primer, a DNA polymerase, radioactively or fluorescently labeled nucleotides, and modified nucleotides that terminate DNA strand elongation. The DNA sample is divided into four separate sequencing reactions, containing all four of the standard deoxynucleotides (dATP, dGTP, dCTP and dTTP) and the DNA polymerase. To each reaction is added only one of the four dideoxynucleotides (ddATP, ddGTP, ddCTP, or ddTTP).

These dideoxynucleotides are the chain-terminating nucleotides, lacking a 3'-OH group required for the formation of a phosphodiester bond between two nucleotides during DNA strand elongation. Incorporation of a dideoxynucleotide into the nascent (elongating) DNA strand therefore terminates DNA strand extension, resulting in various DNA fragments of varying length. The dideoxynucleotides are added at lower concentration than the standard deoxynucleotides to allow strand elongation sufficient for sequence analysis.

The newly synthesized and labeled DNA fragments are heat denatured, and separated by size (with a resolution of just one nucleotide) by gel electrophoresis on a denaturing polyacrylamide-urea gel. Each of the four DNA synthesis reactions is run in one of four individual lanes (lanes A, T, G, C); the DNA bands are then visualized by autoradiography or UV light, and the DNA sequence can be directly read off the X-ray film or gel image. In the image on the right, X-ray film was exposed to the gel, and the dark bands correspond to DNA fragments of different lengths.

A dark band in a lane indicates a DNA fragment that is the result of chain termination after incorporation of a dideoxynucleotide (ddATP, ddGTP, ddCTP, or ddTTP). The terminal nucleotide base can be identified according to which dideoxynucleotide was added in the reaction giving that band. The relative positions of the different bands among the four lanes are then used to read (from bottom to top) the DNA sequence as indicated.

There are some technical variations of chain-termination sequencing. In one method, the DNA fragments are tagged with nucleotides containing radioactive phosphorus for radiolabelling. Alternatively, a primer labeled at the 5′ end with a fluorescent dye is used for the tagging. Four separate reactions are still required, but DNA fragments with dye labels can be read using an optical system, facilitating faster and more economical analysis and automation.

This approach is known as 'dye-primer sequencing'. The later development by L Hood and coworkers of fluorescently labeled ddNTPs and primers set the stage for automated, high-throughput DNA sequencing.

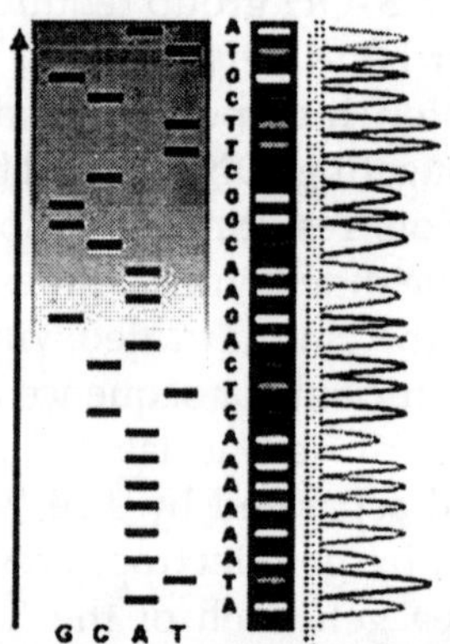

Fig. Sequence Ladder by Radioactive Sequencing Compared to Fluorescent Peaks

The different chain-termination methods have greatly simplified the amount of work and planning needed for DNA sequencing. For example, the chain-termination-based "Sequenase" kit from USB Biochemicals contains most of the reagents needed for sequencing, prealiquoted and ready to use.

Some sequencing problems can occur with the Sanger method, such as non-specific binding of the primer to the DNA, affecting accurate read-out of the DNA sequence. In addition, secondary structures within the DNA template, or contaminating RNA randomly priming at the DNA template can also affect the fidelity of the obtained sequence. Other contaminants affecting the reaction may consist of extraneous DNA or inhibitors of the DNA polymerase.

Dye-terminator Sequencing

An alternative to primer labelling is labelling of the chain terminators, a method commonly called 'dye-terminator sequencing'. The major advantage of this method is that the sequencing can be performed in a single reaction, rather than four reactions as in the labelled-primer method. In dye-terminator sequencing, each of the four dideoxynucleotide chain terminators is labelled with a different fluorescent dye, each fluorescing at a different wavelength.

This method is attractive because of its greater expediency and speed and is now the mainstay in automated sequencing with computer-controlled sequence analyzers. Its potential limitations include dye effects due to differences in the incorporation of the dye-labelled chain terminators into the DNA fragment, resulting in unequal peak heights and shapes in the electronic DNA sequence trace chromatogram after capillary electrophoresis.

This problem has largely been overcome with the introduction of new DNA polymerase enzyme systems and dyes that minimize incorporation variability, as well as methods for eliminating "dye blobs", caused by certain chemical characteristics of the dyes that can result in artifacts in DNA sequence traces. The dye-terminator sequencing method, along with automated high-throughput DNA sequence analyzers, is now being used for the vast majority of sequencing projects, as it is both easier to perform and lower in cost than most previous sequencing methods.

Challenges

Modern sequencing typically produces a sequence that has poor quality in the first 15-40 bases, a high quality region of 700-900 bases, and then quickly deteriorating quality. Base calling software typically outputs an estimate of quality along with the sequence to aid in quality trimming. Before the DNA can be sequenced, linker sequences are attached to its ends, and it is inserted into a cloning vector. The resulting sequence can therefore often contain parts of the vector or the linker sequences, which must be filtered out prior to analysis.

In contrast, emerging sequencing technologies based on pyrosequencing often avoid using cloning vectors. During PCR amplification, unrelated sequences can hybridize, and the resulting clone can be a chimaeric sequence, containing fragments from both sequences. Another problem is polymerase stuttering, where the polymerase repeatedly outputs the same fragments, giving an artificially long low-complexity part of the sequence.

Automation and Sample Preparation

Modern automated DNA sequencing instruments (DNA sequencers) can sequence up to 384 fluorescently labelled samples in a single batch (run) and perform as many as 24 runs a day. However, automated DNA sequencers carry out only DNA-size-based separation (by capillary electrophoresis), detection and recording of dye fluorescence, and data output as fluorescent peak trace chromatograms. Sequencing reactions by thermocycling, cleanup and re-suspension in a buffer solution before loading onto the sequencer are performed separately.

In the past, an operator had to trim the low quality ends of every sequence manually in order to remove the sequencing errors. However, today, software like Chromatogram Explorer or DNA Baser can automatically trim the ends at batch. Such programs score the peaks of each base for quality and remove low-quality base peaks(generally located at the ends of the sequence). If too many bases have a quality score value below a certain threshold, those bases will be automatically deleted. The accuracy of such algorithms is currently below visual examination by a human operator, but is high enough for processing of sequence data sets that are too large for manual examination.

LARGE-SCALE SEQUENCING STRATEGIES

Current methods can directly sequence only relatively short (300-1000 nucleotides long) DNA fragments in a single reaction. The main obstacle to sequencing DNA fragments above this size limit is insufficient power of separation for resolving large DNA fragments that differ in length by only one nucleotide.

Large-scale sequencing aims at sequencing very long DNA fragments. Even relatively small bacterial genomes contain millions of nucleotides, and the human chromosome 1 alone contains about 246 million bases. Therefore, some approaches consist of cutting (with restriction enzymes) or shearing (with mechanical forces) large DNA fragments into shorter DNA fragments. The fragmented DNA is cloned into a DNA vector, usually a bacterial artificial chromosome (BAC), and amplified in *Escherichia coli*.

The amplified DNA can then be purified from the bacterial cells (a disadvantage of bacterial clones for sequencing is that some DNA sequences may be inherently *un-clonable* in some or all available bacterial strains, due to deleterious effect of the cloned sequence on the host bacterium or other effects). These short DNA fragments purified from individual bacterial colonies are then individually and completely sequenced and assembled electronically into one long, contiguous sequence by identifying 100%-identical overlapping sequences between them (shotgun sequencing).

This method does not require any pre-existing information about the sequence of the DNA and is often referred to as *de novo* sequencing. Gaps in the assembled sequence may be filled by Primer walking, often with sub-cloning steps (or transposon-based sequencing depending on the size of the remaining region to be sequenced). These strategies all involve taking many small *reads* of the DNA by one of the above methods and subsequently assembling them into a contiguous sequence. The different strategies have different tradeoffs in speed and accuracy; the shotgun method is the most practical for sequencing large genomes, but its assembly process is complex and potentially error-prone - particularly in the presence of sequence repeats. Because of this, the assembly of the human genome is not literally complete—the repetitive sequences of the centromeres, telomeres, and some other parts of chromosomes result in gaps in the genome assembly.

Despite having only 93% of the full genome assembled, the Human Genome Project was declared complete because their definition of human genome sequencing was limited to euchromatic sequence (99% complete at the time), excluding these intractable repetitive regions. The human genome is about 3 billion (3,000,000,000) bp long; if the average fragment length is 500 bases, it would take a minimum of six million (3 billion/500) to sequence the human genome (not allowing for overlap = 1-fold coverage).

Keeping track of such a high number of sequences presents significant challenges, only held down by developing and coordinating several procedural and computational algorithms, such as efficient database development and management. *Resequencing* or *targeted sequencing* is utilized for determining a change in DNA sequence from a "reference" sequence. It is often performed using PCR to amplify the region of interest (pre-existing DNA sequence is required to design the PCR primers). Resequencing uses three steps, extraction of DNA or RNA from biological tissue; amplification of the RNA or DNA (often by PCR); followed by sequencing. The resultant sequence is compared to a reference or a normal sample to detect mutations.

Chapter 11

Sequencing Methods

The high demand for low cost sequencing has given rise to a number of high-throughput sequencing technologies. These efforts have been funded by public and private institutions as well as privately researched and commercialized by biotechnology companies. High-throughput sequencing technologies are intended to lower the cost of sequencing DNA libraries beyond what is possible with the current dye-terminator method based on DNA separation by capillary electrophoresis.

Many of the new high-throughput methods use methods that parallelize the sequencing process, producing thousands or millions of sequences at once. As molecular detection methods are often not sensitive enough for single molecule sequencing, most approaches use an *in vitro* cloning step to generate many copies of each individual molecule. Emulsion PCR is one method, isolating individual DNA molecules along with primer-coated beads in aqueous bubbles within an oil phase.

A polymerase chain reaction (PCR) then coats each bead with clonal copies of the isolated library molecule and these beads are subsequently immobilized for later sequencing. Emulsion PCR is used in the methods published by Marguilis et al., Shendure and Porreca et al. (also known as "polony sequencing") and SOLID sequencing, (developed by Agencourt and acquired by Applied Biosystems). Another method for *in vitro* clonal amplification is "bridge PCR", where fragments are amplified upon primers attached to a solid surface, developed and used by Solexa (now owned by

Illumina). These methods both produce many physically isolated locations which each contain many copies of a single fragment. The single-molecule method developed by Stephen Quake's laboratory skips this amplification step, directly fixing DNA molecules to a surface.

PARALLELIZED SEQUENCING

Once clonal DNA sequences are physically localized to separate positions on a surface, various sequencing approaches may be used to determine the DNA sequences of all locations, in parallel. "Sequencing by synthesis", like the popular dye-termination electrophoretic sequencing, uses the process of DNA synthesis by DNA polymerase to identify the bases present in the complementary DNA molecule. Reversible terminator methods use reversible versions of dye-terminators, adding one nucleotide at a time, detecting fluorescence corresponding to that position, then removing the blocking group to allow the polymerization of another nucleotide.

Pyrosequencing also uses DNA polymerization to add nucleotides, adding one type of nucleotide at a time, then detecting and quantifying the number of nucleotides added to a given location through the light emitted by the release of attached pyrophosphates. Sequencing by ligation" is another enzymatic method of sequencing, using a DNA ligase enzyme rather than polymerase to identify the target sequence. Used in the polony method and in the solid technology offered by Applied Biosystems, this method uses a pool of all possible oligonucleotides of a fixed length, labeled according to the sequenced position. Oligonucleotides are annealed and ligated; the preferential ligation by DNA ligase for matching sequences results in a signal corresponding to the complementary sequence at that position.

ELISA

Enzyme-Linked ImmunoSorbent Assay, also called ELISA, Enzyme ImmunoAssay or EIA, is a biochemical technique used mainly in immunology to detect the presence of an antibody or an antigen in a sample. The ELISA has been used as a diagnostic tool in medicine and plant pathology, as well as a quality control check in various industries. In simple terms, in ELISA an unknown amount of antigen is affixed to a surface, and then a specific antibody is washed over the surface so that it can bind to the antigen.

This antibody is linked to an enzyme, and in the final step a substance is added that the enzyme can convert to some detectable signal. Thus in the case of fluorescence ELISA, when light is shone upon the sample, any antigen/antibody complexes will fluoresce so that the amount of antigen in the sample can be measured. Performing an ELISA involves at least one antibody with specificity for a particular antigen. The sample with an unknown amount of antigen is immobilized on a solid support (usually a polystyrene microtiter plate) either non-specifically (via adsorption to the surface) or specifically (via capture by another antibody specific to the same antigen, in a "sandwich" ELISA).

After the antigen is immobilized the detection antibody is added, forming a complex with the antigen. The detection antibody can be covalently linked to an enzyme, or can itself be detected by a secondary antibody which is linked to an enzyme through bioconjugation.

Between each step the plate is typically washed with a mild detergent solution to remove any proteins or antibodies that are not specifically bound. After the final wash step the plate is developed by adding an enzymatic substrate to produce a visible signal, which indicates the quantity of antigen in the sample. Older ELISAs utilize chromogenic substrates, though newer assays employ fluorogenic substrates with much higher sensitivity.

APPLICATIONS

Because the ELISA can be performed to evaluate either the presence of antigen or the presence of antibody in a sample, it is a useful tool both for determining serum antibody concentrations (such as with the HIV test or West Nile Virus) and also for detecting the presence of antigen. It has also found applications in the food industry in detecting potential food allergens such as milk, peanuts, walnuts, almonds, and eggs. ELISA can also be used in toxicology as a rapid presumptive screen for certain classes of drugs. The ELISA test, or the enzyme immunoassay (EIA), was the first screening test commonly employed for HIV. It has a high sensitivity. In an ELISA test, a person's serum is diluted 400-fold and applied to a plate to which HIV antigens have been attached. If antibodies to HIV are present in the serum, they may bind to these HIV antigens.

The plate is then washed to remove all other components of the serum. A specially prepared "secondary antibody" — an antibody that binds to other antibodies — is then applied to the plate, followed by another wash. This secondary antibody is chemically linked in advance to an enzyme. Thus the plate will contain enzyme in proportion to the amount of secondary antibody bound to the plate. A substrate for the enzyme is applied, and catalysis by the enzyme leads to a change in colour or fluorescence. ELISA results are reported as a number; the most controversial aspect of this test is determining the "cut-off" point between a positive and negative result. One method of determining a cut-off point is by comparison with a known standard.

For example, if an ELISA test will be used in workplace drug screening, a cut-off concentration (e.g., 50 ng/mL of drug) will be established and a sample will be prepared that contains that concentration of analyte.

Unknowns that generate a signal that is stronger than the known sample are called "positive"; those that generate weaker signal are called "negative." Prior to the development of the EIA/ELISA, the only option for conducting an immunoassay was radioimmu-noassay, a technique using radioactively-labeled antigens or antibodies. In radioimm-unoassay, the radioactivity provides the signal which indicates whether a specific antigen or antibody is present in the sample.

Radioimmunoassay was first described in a paper by Rosalyn Sussman Yalow and Solomon Berson published in 1960. Because radioactivity poses a health threat, a safer alternative was sought. A suitable alternative to radioimmunoassay would substitute a non-radioactive signal in place of the radioactive signal. When certain enzymes (such as peroxidase) react with appropriate substrates (such as ABTS or 3,3',5,5'-Tetramethylbenzidine), they can result in changes in colour, which can be used as a signal.

However, the signal has to be associated with the presence of antibody or antigen, which is why the enzyme has to be linked to an appropriate antibody.

This linking process was independently developed by Stratis Avrameas and G.B. Pierce. Since it is necessary to remove any unbound antibody or antigen by washing, the antibody or antigen has to be fixed to the surface of the container, i.e. the *immunosorbent* has to be prepared. A technique to accomplish this was published by Wide and Porath in 1966.

In 1971, Peter Perlmann and Eva Engvall at Stockholm University in Sweden, as well as Anton Schuurs and Bauke van Weemen in The Netherlands, independently published papers which synthesized this knowledge into methods to perform EIA/ELISA

TYPES

INDIRECT" ELISA

The steps of the general, "indirect," ELISA for determining serum antibody concentrations are:

Apply a sample of known antigen of known concentration to a surface, often the well of a microtiter plate. The antigen is fixed to the surface to render it immobile. Simple adsorption of the protein to the plastic surface is usually sufficient. These samples of known antigen concentrations will constitute a standard curve used to calculate antigen concentrations of unknown samples. Note that the antigen itself may be an antibody.

The plate wells or other surface are then coated with serum samples of unknown antigen concentration, diluted into the same buffer used for the antigen standards. Since antigen immobilization in this step is due to non-specific adsorption, it is important for the total protein concentration to be similar to that of the antigen standards. A concentrated solution of non-interacting protein, such as Bovine Serum Albumin (BSA) or casein, is added to all plate wells. This step is known as blocking, because the serum proteins block non-specific absorption of other proteins to the plate.

The plate is washed, and a detection antibody specific to the antigen of interest is applied to all plate wells. This antibody will only bind to immobilized antigen on the well surface, not to other serum proteins or the blocking proteins.

Secondary antibodies, which will bind to any remaining detection antibodies, are added to the wells. These secondary antibodies are conjugated to the substrate-specific enzyme. This step may be skipped if the detection antibody is conjugated to an enzyme. Wash the plate, so that excess unbound enzyme-antibody conjugates are removed.

Apply a substrate which is converted by the enzyme to elicit a chromogenic or fluorogenic or electrochemical signal. View/quantify the result using a spectrophotometer, spectrofluorometer, or other optical/electrochemical device.

The enzyme acts as an amplifier; even if only few enzyme-linked antibodies remain bound, the enzyme molecules will produce many signal molecules.

A major disadvantage of the indirect ELISA is that the method of antigen immobilization is non-specific; any proteins in the sample will stick to the microtiter plate well, so small concentrations of analyte in serum must compete with other serum proteins when binding to the well surface. The sandwich ELISA provides a solution to this problem.

ELISA may be run in a qualitative or quantitative format. Qualitative results provide a simple positive or negative result for a sample. The cutoff between positive and negative is determined by the analyst and may be statistical.

Two or three times the standard deviation is often used to distinguish positive and negative samples. In quantitative ELISA, the optical density or fluorescent units of the sample is interpolated into a standard curve, which is typically a serial dilution of the target.

SANDWICH ELISA

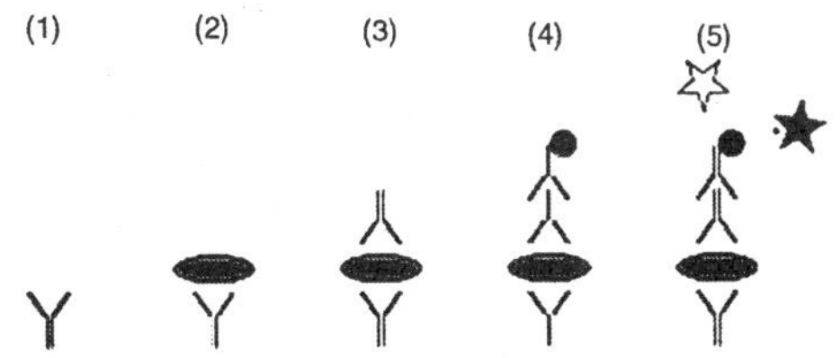

Fig. A Sandwich ELISA

- Plate is coated with a capture antibody;
- Sample is added, and any antigen present binds to capture antibody;
- Detecting antibody is added, and binds to antigen;
- Enzyme-linked secondary antibody is added, and binds to detecting antibody;
- Substrate is added, and is converted by enzyme to detectable form.

A less-common variant of this technique, called "Sandwich" ELISA, is used to detect sample antigen. The steps are as follows:

- Prepare a surface to which a known quantity of capture antibody is bound.
- Block any non specific binding sites on the surface.
- Apply the antigen-containing sample to the plate.
- Wash the plate, so that unbound antigen is removed.
- Apply primary antibodies that bind specifically to the antigen.
- Apply enzyme-linked secondary antibodies which are specific to the primary antibodies.
- Wash the plate, so that the unbound antibody-enzyme conjugates are removed.
- Apply a chemical which is converted by the enzyme into a colour or fluorescent or electrochemical signal.
- Measure the absorbance or fluorescence or electrochemical signal (e.g., current) of the plate wells to determine the presence and quantity of antigen.

The image to the right includes the use of a secondary antibody conjugated to an enzyme, though technically this is not necessary if the primary antibody is conjugated to an enzyme. However, use of a secondary-antibody conjugate avoids the expensive process of creating enzyme-linked antibodies for every antigen one might want to detect. By using an enzyme-linked antibody that binds the Fc region of other antibodies, this same enzyme-linked antibody can be used in a variety of situations.

The major advantage of a sandwich ELISA is the ability to use crude or impure samples and still selectively bind any antigen that may be present. Without the first layer of "capture" antibody, any proteins in the sample (including serum proteins) may competitively adsorb to the plate surface, lowering the quantity of antigen immobilized.

COMPETITIVE ELISA

A third use of ELISA is through competitive binding. The steps for this ELISA are somewhat different than the first two examples:

- Unlabeled antibody is incubated in the presence of its antigen.
- These bound antibody/antigen complexes are then added to an antigen coated well.
- The plate is washed, so that unbound antibody is removed. (The more antigen in the sample, the less antibody will be able to bind to the antigen in the well, hence "competition.")
- The secondary antibody, specific to the primary antibody is added. This second antibody is coupled to the enzyme.
- A substrate is added, and remaining enzymes elicit a chromogenic or fluorescent signal.

For competitive ELISA, the higher the original antigen concentration, the weaker the eventual signal.

(Note that some competitive ELISA kits include enzyme-linked antigen rather than enzyme-linked antibody. The labeled antigen competes for primary antibody binding sites with your sample antigen (unlabeled). The more antigen in the sample, the less labeled antigen is retained in the well and the weaker the signal).

ELISA REVERSE METHOD & DEVICE

A newer technique uses an solid phase made up of an immunosorbent polystyrene rod with 4-12 protruding ogives. The entire device is immersed in a test tube containing the collected sample and the following steps (washing, incubation in conjugate and incubation in chromogenous) are carried out by dipping the ogives in microwells of standard microplates pre-filled with reagents.

Advantages:

- The ogives can each be sensitized to a different reagent, allowing the simultaneous detection of different antibodies and different antigens for multi-target assays;
- The sample volume can be increased to improve the test sensitivity in clinical (saliva, urine), food (bulk milk, pooled eggs) and environmental (water) samples;
- One ogive is left unsensitized to measure the non-specific reactions of the sample;
- The use of laboratory supplies for dispensing sample aliquots, washing solution and reagents in microwells is not required, facilitating ready-to-use lab-kits and on-site kits.

NORTHERN BLOT

The northern blot is a technique used in molecular biology research to study gene expression. It takes its name from its similarity to the Southern blot technique, named for biologist Edwin Southern and used to study DNA. The major difference is that RNA, rather than DNA, is analyzed in the northern blot. Both techniques use electrophoresis and detection with a hybridization probe. The northern blot technique was developed in 1977 by James Alwine, David Kemp, and George Stark at Stanford University.

The gels may be run on either agarose or denaturing polyacrylamide, the later being preferable for smaller fragments of RNA. Unlike in the Southern blot, formaldehyde is added to the gel and acts as a denaturant to agarose. For polyacrylamide, urea is the denaturant. As in the Southern blot, the hybridization probe may be made from DNA or RNA. A variant of the procedure known as the reverse northern blot is occasionally used. In this procedure, the substrate nucleic acid (that is affixed to the membrane) is a collection of isolated DNA fragments, and the probe is RNA extracted from a tissue and radioactively labelled.

The use of DNA microarrays that have come into widespread use in the late 1990s and early 2000s is more akin to the reverse procedure, in that they involve the use of isolated DNA fragments affixed to a substrate, and hybridization with a probe made from cellular RNA. Thus the reverse procedure, though originally uncommon, enabled northern analysis to evolve into gene expression profiling, in which many (possibly all) of the genes in an organism may have their expression monitored.

SOUTHERN BLOT

A Southern blot is a method routinely used in molecular biology to check for the presence of a DNA sequence in a DNA sample. Southern blotting combines agarose gel electrophoresis for size separation of DNA with methods to transfer the size-separated DNA to a filter membrane for probe hybridization. The method is named after its inventor, the British biologist Edwin Southern. Other blotting methods (i.e., western blot, northern blot, southwestern blot) that employ similar principles, but using RNA or protein, have later been named in reference to Southern's name. As the technique was eponymously named, Southern blot should be capitalized, whereas northern and western blots should not.

Method

- Restriction endonucleases are used to cut high-molecular-weight DNA strands into smaller fragments.
- The DNA fragments are then electrophoresed on an agarose gel to separate them by size.
- If some of the DNA fragments are larger than 15 kb, then prior to blotting, the gel may be treated with

an acid, such as dilute HCl, which depurinates the DNA fragments, breaking the DNA into smaller pieces, thus allowing more efficient transfer from the gel to membrane.

- If alkaline transfer methods are used, the DNA gel is placed into an alkaline solution (typically containing sodium hydroxide) to denature the double-stranded DNA. The denaturation in an alkaline environment provides for improved binding of the negatively charged DNA to a positively charged membrane, separates it into single DNA strands for later hybridization to the probe, and destroys any residual RNA that may still be present in the DNA.
- A sheet of nitrocellulose (or, alternatively, nylon) membrane is placed on top of the gel. Pressure is applied evenly to the gel (either using suction, or by placing a stack of paper towels and a weight on top of the membrane and gel), to ensure good and even contact between gel and membrane. Buffer transfer by capillary action from a region of high water potential to a region of low water potential (usually filter paper and paper tissues) is then used to move the DNA from the gel on to the membrane; ion exchange interactions bind the DNA to the membrane due to the negative charge of the DNA and positive charge of the membrane.
- The membrane is then baked, i.e., exposed to high temperature (60 to 100 °C) (in the case of nitrocellulose) or exposed to ultraviolet radiation (nylon) to permanently and covalently crosslink the DNA to the membrane.
- The membrane is then exposed to a hybridization probe—a single DNA fragment with a specific sequence whose presence in the target DNA is to be determined. The probe DNA is labelled so that it can be detected, usually by incorporating radioactivity or tagging the molecule with a fluorescent or

chromogenic dye. In some cases, the hybridization probe may be made from RNA, rather than DNA. To ensure the specificity of the binding of the probe to the sample DNA, most common hybridization methods use salmon testes (sperm) DNA for blocking of the membrane surface and target DNA, deionized formamide, and detergents such as SDS to reduce non-specific binding of the probe.

- After hybridization, excess probe is washed from the membrane, and the pattern of hybridization is visualized on X-ray film by autoradiography in the case of a radioactive or fluorescent probe, or by development of colour on the membrane if a chromogenic detection method is used.

Hybridization of the probe to a specific DNA fragment on the filter membrane indicates that this fragment contains DNA sequence that is complementary to the probe.

The transfer step of the DNA from the electrophoresis gel to a membrane permits easy binding of the labeled hybridization probe to the size-fractionated DNA. It also allows for the fixation of the target-probe hybrids, required for analysis by autoradiography or other detection methods.

Southern blots performed with restriction enzyme-digested genomic DNA may be used to determine the number of sequences (e.g., gene copies) in a genome. A probe that hybridizes only to a single DNA segment that has not been cut by the restriction enzyme will produce a single band on a Southern blot, whereas multiple bands will likely be observed when the probe hybridizes to several highly similar sequences (e.g., those that may be the result of sequence duplication).

Modification of the hybridization conditions (for example, increasing the hybridization temperature or decreasing salt concentration) may be used to increase specificity and decrease hybridization of the probe to sequences that are less than 100% similar.

WESTERN BLOT

The western blot (alternately, immunoblot) is a method

of detecting specific proteins in a given sample of tissue homogenate or extract. It uses gel electrophoresis to separate native or denatured proteins by the length of the polypeptide (denaturing conditions) or by the 3-D structure of the protein (native/ non-denaturing conditions).

The proteins are then transferred to a membrane (typically nitrocellulose or PVDF), where they are probed (detected) using antibodies specific to the target protein. There are now many reagent companies that specialize in providing antibodies (both monoclonal and polyclonal antibodies) against many thousands of different proteins. Commercial antibodies can be expensive, though the unbound antibody can be reused between experiments. This method is used in the fields of molecular biology, biochemistry, immunogenetics and other molecular biology disciplines.

Other related techniques include using antibodies to detect proteins in tissues and cells by immunostaining and enzyme-linked immunosorbent assay (ELISA). The method originated from the laboratory of George Stark at Stanford. The name western blot was given to the technique by W. Neal Burnette and is a play on the name Southern blot, a technique for DNA detection developed earlier by Edwin Southern. Detection of RNA is termed northern blotting.

STEPS IN A WESTERN BLOT

TISSUE PREPARATION

Samples may be taken from whole tissue or from cell culture. In most cases, solid tissues are first broken down mechanically using a blender (for larger sample volumes), using a homogenizer (smaller volumes), or by sonication. Cells may also be broken open by one of the above mechanical methods. However, it should be noted that bacteria, virus or environmental samples can be the source of protein and thus Western blotting is not restricted to cellular studies only.

Assorted detergents, salts, and buffers may be employed to encourage lysis of cells and to solubilize proteins. Protease and phosphatase inhibitors are often added to prevent the digestion of the sample by its own enzymes. A combination of biochemical and mechanical techniques – including various types of filtration and centrifugation – can be used to separate different cell compartments and organelles.

GEL ELECTROPHORESIS

The proteins of the sample are separated using gel electrophoresis. Separation of proteins may be by isoelectric point (pI), molecular weight, electric charge, or a combination of these factors. The nature of the separation depends on the treatment of the sample and the nature of the gel.

By far the most common type of gel electrophoresis employs polyacrylamide gels and buffers loaded with sodium dodecyl sulfate (SDS). SDS-PAGE (SDS polyacrylamide gel electrophoresis) maintains polypeptides in a denatured state once they have been treated with strong reducing agents to remove secondary and tertiary structure (e.g. S-S disulfide bonds to SH and SH) and thus allows separation of proteins by their molecular weight.

Sampled proteins become covered in the negatively charged SDS and move to the positively charged electrode through the acrylamide mesh of the gel. Smaller proteins migrate faster through this mesh and the proteins are thus separated according to size (usually measured in kilo Daltons, kD).

The concentration of acrylamide determines the resolution of the gel - the greater the acrylamide concentration the better the resolution of lower molecular weight proteins. The lower the acrylamide concentration the better the resolution of higher molecular weight proteins. Proteins travel only in one dimension along the gel for most blots.

Samples are loaded into *wells* in the gel. One lane is usually reserved for a *marker* or *ladder*, a commercially available mixture of proteins having defined molecular weights, typically stained so as to form visible, coloured bands. An example of a ladder is the GE Full Range Molecular weight ladder. When voltage is applied along the gel, proteins migrate into it at different speeds. These different rates of advancement (different *electrophoretic mobilities*) separate into *bands* within each *lane.*

It is also possible to use a two-dimensional (2-D) gel which spreads the proteins from a single sample out in two dimensions. Proteins are separated according to isoelectric point (pH at which they have neutral net charge) in the first dimension, and according to their molecular weight in the second dimension.

TRANSFER

In order to make the proteins accessible to antibody detection, they are moved from within the gel onto a membrane made of nitrocellulose or polyvinylidene fluoride (PVDF). The membrane is placed on top of the gel, and a stack of tissue papers placed on top of that. The entire stack is placed in a buffer solution which moves up the paper by capillary action, bringing the proteins with it. Another method for transferring the proteins is called electroblotting and uses an electric current to pull proteins from the gel into the PVDF or nitrocellulose membrane.

The proteins move from within the gel onto the membrane while maintaining the organization they had within the gel. As a result of this "blotting" process, the proteins are exposed on a thin surface layer for detection. Both varieties of membrane are chosen for their non-specific protein binding properties (i.e. binds all proteins equally well). Protein binding is based upon hydrophobic interactions, as well as charged interactions between the membrane and protein. Nitrocellulose membranes are cheaper than PVDF, but are far more fragile and do not stand up well to repeated probings.

The uniformity and overall effectiveness of transfer of protein from the gel to the membrane can be checked by staining the membrane with Coomassie or Ponceau S dyes. Coomassie is the more sensitive of the two, although Ponceau S's water solubility makes it easier to subsequently destain and probe the membrane as described below.

BLOCKING

Since the membrane has been chosen for its ability to bind

protein, and both antibodies and the target are proteins, steps must be taken to prevent interactions between the membrane and the antibody used for detection of the target protein. Blocking of non-specific binding is achieved by placing the membrane in a dilute solution of protein - typically Bovine serum albumin (BSA) or non-fat dry milk (both are inexpensive), with a minute percentage of detergent such as Tween 20.

The protein in the dilute solution attaches to the membrane in all places where the target proteins have not attached. Thus, when the antibody is added, there is no room on the membrane for it to attach other than on the binding sites of the specific target protein. This reduces "noise" in the final product of the Western blot, leading to clearer results, and eliminates false positives.

DETECTION

During the detection process the membrane is "probed" for the protein of interest with a modified antibody which is linked to a reporter enzyme, which when exposed to an appropriate substrate drives a colourimetric reaction and produces a colour. For a variety of reasons, this traditionally takes place in a two-step process, although there are now one-step detection methods available for certain applications.

Two step

Primary antibody

Antibodies are generated when a host species or immune cell culture is exposed to the protein of interest (or a part thereof). Normally, this is part of the immune response, whereas here they are harvested and used as sensitive and specific detection tools that bind the protein directly.

After blocking, a dilute solution of primary antibody (generally between 0.5 and 5 micrograms/ml) is incubated with the membrane under gentle agitation. Typically, the solution is composed of buffered saline solution with a small percentage of detergent, and sometimes with powdered milk or BSA. The antibody solution and the membrane can be sealed and incubated together for anywhere from 30 minutes to overnight. It can also be incubated at different temperatures, with warmer temperatures being associated with more binding, both specific (to the target protein, the "signal") and non-specific ("noise").

Secondary antibody

After rinsing the membrane to remove unbound primary antibody, the membrane is exposed to another antibody, directed at a species-specific portion of the primary antibody. This is known as a secondary antibody, and due to its targeting properties, tends to be referred to as "anti-mouse," "anti-goat," etc. Antibodies come from animal sources (or animal sourced hybridoma cultures); an anti-mouse secondary will bind to just about any mouse-sourced primary antibody.

This allows some cost savings by allowing an entire lab to share a single source of mass-produced antibody, and provides far more consistent results. The secondary antibody is usually linked to biotin or to a reporter enzyme such as alkaline phosphatase or horseradish peroxidase. This means that several secondary antibodies will bind to one primary antibody and enhances the signal.

Most commonly, a horseradish peroxidase-linked secondary is used in conjunction with a chemiluminescent agent, and the reaction product produces luminescence in proportion to the amount of protein. A sensitive sheet of photographic film is placed against the membrane, and exposure to the light from the reaction creates an image of the antibodies bound to the blot.

As with the ELISPOT and ELISA procedures, the enzyme can be provided with a substrate molecule that will be converted by the enzyme to a colored reaction product that will be visible on the membrane.

A third alternative is to use a radioactive label rather than an enzyme coupled to the secondary antibody, such as labeling an antibody-binding protein like *Staphylococcus* Protein A with a radioactive isotope of iodine. Since other methods are safer, quicker and cheaper this method is now rarely used.

ONE STEP

Historically, the probing process was performed in two steps because of the relative ease of producing primary and secondary antibodies in separate processes. This gives researchers and corporations huge advantages in terms of flexibility, and adds an amplification step to the detection process. Given the advent of high-throughput protein analysis and lower limits of detection, however, there has been interest in developing one-step probing systems that would allow the process to occur faster and with less consumables.

This requires a probe antibody which both recognizes the protein of interest and contains a detectable label, probes which are often available for known protein tags. The primary probe is incubated with the membrane in a manner similar to that for the primary antibody in a two-step process, and then is ready for direct detection after a series of wash steps.

ANALYSIS

After the unbound probes are washed away, the Western blot is ready for detection of the probes that are labeled and bound to the protein of interest. In practical terms, not all Westerns reveal protein only at one band in a membrane. Size approximations are taken by comparing the stained bands to that of the marker or ladder loaded during electrophoresis.

The process is repeated for a structural protein, such as actin or tubulin, that should not change between samples. The amount of target protein is indexed to the structural protein to control between groups. This practice ensures correction for the amount of total protein on the membrane in case of errors or incomplete transfers.

COLORIMETRIC DETECTION

The colorimetric detection method depends on incubation of the Western blot with a substrate that reacts with the reporter enzyme (such as peroxidase) that is bound to the secondary antibody. This converts the soluble dye into an insoluble form of a different colour that precipitates next to the enzyme and thereby stains the membrane. Development of the blot is then stopped by washing away the soluble dye. Protein levels are evaluated through densitometry or spectrophotometry.

CHEMILUMINESCENCE

Chemiluminescent detection methods depend on incubation of the Western blot with a substrate that will luminesce when exposed to the reporter on the secondary antibody. The light is then detected by photographic film, and more recently by CCD cameras which captures a digital image of the Western blot.

The image is analysed by densitometry, which evaluates the relative amount of protein staining and quantifies the results in terms of optical density. Newer software allows further data analysis such as molecular weight analysis if appropriate standards are used. So-called "enhanced chemiluminescent" (ECL) detection is considered to be among the most sensitive detection methods for blotting analysis.

RADIOACTIVE DETECTION

Radioactive labels do not require enzyme substrates, but rather allow the placement of medical X-ray film directly against the Western blot which develops as it is exposed to the label and creates dark regions which correspond to the protein bands of interest. The importance of radioactive detections methods is declining, because it is very expensive, health and safety risks are high and ECL provides a useful alternative.

FLUORESCENT DETECTION

The fluorescently labeled probe is excited by light and the emission of the excitation is then detected by a photosensor such as CCD camera equipped with appropriate emission filters which captures a digital image of the Western blot and allows further data analysis such as molecular weight analysis and a quantitative Western blot analysis. Fluorescence is considered to be among the most sensitive detection methods for blotting analysis.

SECONDARY PROBING

One major difference between nitrocellulose and PVDF membranes relates to the ability of each to support "stripping" antibodies off and reusing the membrane for subsequent antibody probes. While there are well-established protocols available for stripping nitrocellulose membranes, the sturdier PVDF allows for easier stripping, and for more reuse before background noise limits experiments. Another difference is that, unlike nitrocellulose, PVDF must be soaked in 95% ethanol, isopropanol or methanol before use. PVDF membranes also tend to be thicker and more resistant to damage during use.

2-D GEL ELECTROPHORESIS

2-dimensional SDS-PAGE uses the principles and techniques outlined above. 2-D SDS-PAGE, as the name suggests, involves the migration of polypeptides in 2 dimensions. For example, in the first dimension polypeptides are separated according to isoelectric point, while in the second dimension polypeptides are separated according to their molecular weight. The isoelectric point of a given protein is determined by the relative number of positively (e.g. lysine and arginine) and negatively (e.g. glutamate and aspartate) charged amino acids, with negatively charged amino acids contributing to a high isoelectric point and positively charged amino acids contributing to a low isoelectric point.

Samples could also be separated first under nonreducing conditions using SDS-PAGE and under reducing conditions in the second dimension, which breaks apart disulfide bonds that hold subunits together. SDS-PAGE might also be coupled with urea-PAGE for a 2-dimensional gel. In principle, this method allows for the separation of all cellular proteins on a single large gel. A major advantage of this method is that it often distinguishes between different isoforms of a particular protein - e.g. a protein that has been phosphorylated (by addition of a negatively charged group). Proteins that have been separated can be cut out of the gel and then analysed by mass spectrometry, which identifies the protein.

MEDICAL DIAGNOSTIC APPLICATIONS

- The confirmatory HIV test employs a Western blot to detect anti-HIV antibody in a human serum sample. Proteins from known HIV-infected cells are separated and blotted on a membrane as above. Then, the serum to be tested is applied in the primary antibody incubation step; free antibody is washed away, and a secondary anti-human antibody linked to an enzyme signal is added. The stained bands then indicate the proteins to which the patient's serum contains antibody.
- A Western blot is also used as the definitive test for Bovine spongiform encephalopathy (BSE, commonly referred to as 'mad cow disease').
- Some forms of Lyme disease testing employ Western blotting.

SOUTHWESTERN BLOT

Southwestern blotting, based along the lines of Southern blotting (which was created by Edwin Southern) and first described by B. Bowen and colleagues in 1980, involves identifying and characterizing DNA-binding proteins (proteins that bind to DNA) by their ability to bind to specific oligonucleotide probes. The proteins undergo gel electrophoresis and are subsequently transferred to nitrocellulose membranes similar to other types of blotting.

The name southwestern blotting is based on the fact that this technique detects DNA-binding proteins, since DNA detection is by Southern blotting and protein detection is by western blotting. However, since the first southwestern blottings, many more have been proposed and discovered. Large amounts of proteins and their degradation when being isolated hampered previous protocols.

"Southwestern blot mapping" is performed for rapid characterization of both DNA-binding proteins and their specific sites on genomic DNA. Proteins are separated on a Sodium Dodecyl Sulfate (SDS), polyacrylamide gel (PAGE), renatured by removing SDS in the presence of urea, and blotted onto nitrocellulose by diffusion. The genomic DNA region of interest is digested by restriction enzymes selected to produce fragments of appropriate but different sizes, which are subsequently end-labeled and allowed to bind to the separated proteins. The specifically bound DNA is eluted from each individual protein-DNA complex and analyzed by acrylamide gel electrophoresis.

Evidence that tissue-specific DNA binding proteins may be detected by this technique is presented. Moreover, their sequence-specific binding allows the purification of the corresponding selectively bound DNA fragments and may improve protein-mediated cloning of DNA regulatory sequences.

DNA MICROARRAY

A DNA microarray is a high-throughput technology used in molecular biology and in medicine. It consists of an arrayed series of thousands of microscopic spots of DNA oligonucleotides, called features, each containing picomoles of a specific DNA sequence. This can be a short section of a gene or other DNA element that are used as probes to hybridize a cDNA or cRNA sample (called target) under high-stringency conditions. Probe-target hybridization is usually detected and quantified by fluorescence-based detection of fluorophore-labeled targets to determine relative abundance of nucleic acid sequences in the target.

In standard microarrays, the probes are attached to a solid surface by a covalent bond to a chemical matrix (via epoxy-silane, amino-silane, lysine, polyacrylamide or others). The solid surface can be glass or a silicon chip, in which case they are commonly known as *gene chip* or colloquially *Affy chip* when an Affymetrix chip is used. Other microarray platforms, such as Illumina, use microscopic beads, instead of the large solid support. DNA arrays are different from other types of microarray only in that they either measure DNA or use DNA as part of its detection system.

DNA microarrays can be used to measure changes in expression levels or to detect single nucleotide polymorphisms

(SNPs). Microarrays also differ in fabrication, workings, accuracy, efficiency, and cost. Additional factors for microarray experiments are the experimental design and the methods of analyzing the data.

Microarray technology evolved from Southern blotting, where fragmented DNA is attached to a substrate and then probed with a known gene or fragment. The use of a collection of distinct DNAs in arrays for expression profiling was first described in 1987, and the arrayed DNAs were used to identify genes whose expression is modulated by interferon. These early gene arrays were made by spotting cDNAs onto filter paper with a pin-spotting device.

FABRICATION

Microarrays can be manufactured in different ways, depending on the number of probes under examination, costs, customization requirements, and the type of scientific question being asked. Arrays may have as few as 10 probes to up to 2.1 million micron-scale probes from commercial vendors.

Spotted vs. Oligonucleotide Arrays

Microarrays can be fabricated using a variety of technologies, including printing with fine-pointed pins onto glass slides, photolithography using pre-made masks, photolithography using dynamic micromirror devices, ink-jet printing, or electrochemistry on microelectrode arrays.

In *spotted microarrays*, the probes are oligonucleotides, cDNA or small fragments of PCR products that correspond to mRNAs. The probes are synthesized prior to deposition on the array surface and are then "spotted" onto glass. A common approach utilizes an array of fine pins or needles controlled by a robotic arm that is dipped into wells containing DNA probes and then depositing each probe at designated locations on the array surface. The resulting "grid" of probes represents the nucleic acid profiles of the prepared probes and is ready to receive complementary cDNA or cRNA "targets" derived from experimental or clinical samples. This technique is used by research scientists around the world to produce "in-house" printed microarrays from their own labs. These arrays may be easily customized for each experiment, because researchers can choose the probes and printing locations on the arrays, synthesize the probes in their own lab (or collaborating facility), and spot the arrays.

They can then generate their own labeled samples for hybridization, hybridize the samples to the array, and finally scan the arrays with their own equipment. This provides a relatively low-cost microarray that is customized for each study, and avoids the costs of purchasing often more expensive commercial arrays that may represent vast numbers of genes that are not of interest to the investigator.

Publications exist which indicate in-house spotted microarrays may not provide the same level of sensitivity compared to commercial oligonucleotide arrays, possibly owing to the small batch sizes and reduced printing efficiencies when compared to industrial manufactures of oligo arrays. Applied Microarrays offers a commercial array platform called the "CodeLink" system where 30-mer oligonucleotide probes (sequences of 30 nucleotides in length) are piezoelectrically deposited on an acrylamide matrix without any contact being made between the depositing equipment and the array surface itself. These arrays are comparable in quality to most manufactured arrays and generally superior to in-house printed arrays.

In *oligonucleotide microarrays*, the probes are short sequences designed to match parts of the sequence of known or predicted open reading frames. Although oligonucleotide probes are often used in "spotted" microarrays, the term "oligonucleotide array" most often refers to a specific technique of manufacturing. Oligonucleotide arrays are produced by printing short oligonucleotide sequences designed to represent a single gene or family of gene splice-variants by synthesizing this sequence directly onto the array surface instead of depositing intact sequences.

Sequences may be longer (60-mer probes such as the Agilent design) or shorter (25-mer probes produced by Affymetrix) depending on the desired purpose; longer probes are more specific to individual target genes, shorter probes may be spotted in higher density across the array and are cheaper to manufacture.

One technique used to produce oligonucleotide arrays include photolithographic synthesis (Agilent and Affymetrix) on a silica substrate where light and light-sensitive masking agents are used to "build" a sequence one nucleotide at a time across the entire array. Each applicable probe is selectively "unmasked" prior to bathing the array in a solution of a single nucleotide, then a masking reaction takes place and the next set of probes are unmasked in preparation for a different nucleotide exposure. After many repetitions, the sequences of every probe become fully constructed. More recently, Maskless Array Synthesis from NimbleGen Systems has combined flexibility with large numbers of probes.

TWO-CHANNEL VS. ONE-CHANNEL DETECTION

Two-colour microarrays or *two-channel microarrays* are typically hybridized with cDNA prepared from two samples to be compared (e.g. diseased tissue versus healthy tissue) and that are labeled with two different fluorophores. Fluorescent dyes commonly used for cDNA labelling include Cy3, which has a fluorescence emission wavelength of 570 nm (corresponding to the green part of the light spectrum), and Cy5 with a fluorescence emission wavelength of 670 nm (corresponding to the red part of the light spectrum).

The two Cy-labelled cDNA samples are mixed and hybridized to a single microarray that is then scanned in a microarray scanner to visualize fluorescence of the two fluorophores after excitation with a laser beam of a defined wavelength. Relative intensities of each fluorophore may then be used in ratio-based analysis to identify up-regulated and down-regulated genes.

Oligonucleotide microarrays often contain control probes designed to hybridize with RNA spike-ins. The degree of hybridization between the spike-ins and the control probes is used to normalize the hybridization measurements for the target probes. Although absolute levels of gene expression may be determined in the two-colour array, the relative differences in expression among different spots within a sample and between samples is the preferred method of data analysis for the two-colour system. Examples of providers for such microarrays includes Agilent with their Dual-Mode platform, Eppendorf with their DualChip platform for fluorescence labeling, and TeleChem International with Arrayit.

In *single-channel microarrays* or *one-colour microarrays,* the arrays are designed to give estimations of the absolute levels of gene expression. Therefore the comparison of two conditions requires two separate single-dye hybridizations. As only a single dye is used, the data collected represent absolute values of gene expression.

These may be compared to other genes within a sample or to reference "normalizing" probes used to calibrate data across the entire array and across multiple arrays. Three popular single-channel systems are the Affymetrix "Gene Chip", the Applied Microarrays "CodeLink" arrays, and the Eppendorf "DualChip & Silverquant".

One strength of the single-dye system lies in the fact that an aberrant sample cannot affect the raw data derived from other samples, because each array chip is exposed to only one sample (as opposed to a two-colour system in which a single low-quality sample may drastically impinge on overall data precision even if the other sample was of high quality). Another benefit is that data are more easily compared to arrays from different experiments; the absolute values of gene expression may be compared between studies conducted months or years apart. A drawback to the one-colour system is that, when compared to the two-colour system, twice as many microarrays are needed to compare samples within an experiment.

MICROARRAYS AND BIOINFORMATICS

The advent of inexpensive microarray experiments created several specific bioinformatics challenges:

- The multiple levels of replication in experimental design (Experimental Design)

- The number of platforms and independent groups and data format (Standardization)
- The treatment of the data (Statistical Analysis)
- What exactly are we measuring (Relation between probe and gene)
- The sheer volume of data and the ability to share it (Data Warehousing)

Experimental Design

Due to the biological complexity of gene expression, the considerations of experimental design that are discussed in the expression profiling article are of critical importance if statistically and biologically valid conclusions are to be drawn from the data.

There are three main elements to consider when designing a microarray experiment. First, replication of the biological samples is essential for drawing conclusions from the experiment. Second, technical replicates (two RNA samples obtained from each experimental unit) help to ensure precision and allow for testing differences within treatment groups. The technical replicates may be two independent RNA extractions or two aliquots of the same extraction.

Third, spots of each cDNA clone or oligonucleotide are present as replicates (at least duplicates) on the microarray slide, to provide a measure of technical precision in each hybridization. It is critical that information about the sample preparation and handling is discussed, in order to help identify the independent units in the experiment and to avoid inflated estimates of statistical significance.

Standardization

Microarray data is difficult to exchange due to the lack of standardization in arrays. This presents an interoperability problem in bioinformatics. Various grass-roots open-source projects are trying to ease the exchange and analysis of data produced with non-proprietary chips:

- For example, the "Minimum Information About a Microarray Experiment" (MIAME) checklist helps

define the level of detail that should exist and is being adopted by many journals as a requirement for the submission of papers incorporating microarray results. But MIAME does not describe the format for the information, so while many formats can support the MIAME requirements, as of 2007 no format permits verification of complete semantic compliance.
- The "MicroArray Quality Control (MAQC) Project" is being conducted by the US Food and Drug Administration (FDA) to develop standards and quality control metrics which will eventually allow the use of MicroArray data in drug discovery, clinical practice and regulatory decision-making.
- The MicroArray and Gene Expression Data (MGED) group is working on the standardization of the representation of gene expression data and relevant annotations.

Statistical Analysis

The analysis of DNA microarrays poses a large number of statistical problems, including the normalization of the data. There are several normalization methods in the published literature some of which are platform specific; as in many other cases where authorities disagree, a sound conservative approach is to directly compare different normalization methods to determine the effects of these different methods on the results obtained. This can be done, for example, by investigating the performance of various methods on data from "spike-in" experiments.

Also, experimenters must account for multiple comparisons: even if the statistical P-value assigned to a gene indicates that it is extremely unlikely that differential expression of this gene was due to random rather than treatment effects, the very high number of genes on an array makes it likely that differential expression of some genes represent false positives or false negatives. Statistical methods tailored to microarray analyses have recently become available that assess statistical power based on the variation present in the data and the number of experimental replicates, and can help minimize type I and type II errors in the analyses.

A basic difference between microarray data analysis and much traditional biomedical research is the dimensionality of the data. A large clinical study might collect 100 data items per patient for thousands of patients. A medium-size microarray study will obtain many thousands of numbers per sample for perhaps a hundred samples. Many analysis techniques treat each sample as a single point in a space with thousands of dimensions, then attempt by various techniques to reduce the dimensionality of the data to something humans can visualize.

Relation between probe and gene

The relation between a probe and the mRNA that it is expected to detect is problematic. On the one hand, some mRNAs may cross-hybridize probes in the array that are supposed to detect another mRNA. On the other hand, probes that are designed to detect the mRNA of a particular gene may be relying on genomic EST information that is incorrectly associated with that gene.

Data Warehousing

Microarray data was found to be more useful when compared to other similar datasets. The sheer volume (in bytes), specialized formats (such as MIAME), and curation efforts associated with the datasets require specialized databases to store the data.

Chapter 12

Chromatography

Chromatography the collective term for a family of laboratory techniques for the separation of mixtures. It involves passing a mixture dissolved in a "mobile phase" through a *stationary phase,* which separates the analyte to be measured from other molecules in the mixture and allows it to be isolated.

Chromatography may be preparative or analytical. Preparative chromatography seeks to separate the components of a mixture for further use (and is thus a form of purification). Analytical chromatography normally operates with smaller amounts of material and seeks to measure the relative proportions of analytes in a mixture. The two are not mutually exclusive.

EXPLANATION

An analogy which is sometimes useful is to suppose a mixture of bees and wasps passing over a flower bed. The bees would be more attracted to the flowers than the wasps, and would become separated from them. If one were to observe at a point past the flower bed, the wasps would pass first, followed by the bees. In this analogy, the bees and wasps represent the analytes to be separated, the flowers represent the stationary phase, and the mobile phase could be thought of as the air.

The key to the separation is the differing affinities among analyte, stationary phase, and mobile phase. The observer could represent the detector used in some forms of analytical chromatography. A key point is that the detector need not be capable of discriminating between the analytes, since they have become separated before passing the detector.

The history of chromatography spans from the mid-19th century the 21st. Chromatography, literally "colour writing", was used—and named— in the first decade of the 20th century, primarily for the separation of plant pigments such as chlorophyll. New forms of chromatography developed in the 1930s and 1940s made the technique useful for a wide range of separation processes and chemical analysis tasks, especially in biochemistry. Some related techniques were developed in the 19th century (and even before), but the first true chromatography is usually attributed to Russian botanist Mikhail Semyonovich Tsvet, who used columns of calcium carbonate for separating plant pigments in the first decade of the 20th century during his research on chlorophyll.

Chromatography began to take its modern form following the work of Archer John Porter Martin and Richard Laurence Millington Synge in the 1940s and 1950s. They laid out the principles and basic techniques of partition chromatography, and their work spurred the rapid development of several lines of chromatography methods: paper chromatography, gas chromatography, and what would become known as high performance liquid chromatography.

Since then, the technology has advanced rapidly. Researchers found that the principles underlying Tsvet's chromatography could be applied in many different ways, giving rise to the different varieties of chromatography described below. Simultaneously, advances continually improved the technical performance of chromatography, allowing the separation of increasingly similar molecules.

CHROMATOGRAPHY TERMS

- The analyte is the substance that is to be separated during chromatography.
- Analytical chromatography is used to determine the existence and possibly also the concentration of analyte(s) in a sample.
- A bonded phase is a stationary phase that is

covalently bonded to the support particles or to the inside wall of the column tubing.

- A chromatogram is the visual output of the chromatograph. In the case of an optimal separation, different peaks or patterns on the chromatogram correspond to different components of the separated mixture.

Plotted on the x-axis is the retention time and plotted on the y-axis a signal (for example obtained by a spectrophotometer, mass spectrometer or a variety of other detectors) corresponding to the response created by the analytes exiting the system. In the case of an optimal system the signal is proportional to the concentration of the specific analyte separated.

- A chromatograph is equipment that enables a sophisticated separation e.g. gas chromatographic or liquid chromatographic separation.
- Chromatography is a physical method of separation in which the components to be separated are distributed between two phases, one of which is stationary (stationary phase) while the other (the mobile phase) moves in a definite direction.
- The effluent is the mobile phase leaving the column.
- An immobilized phase is a stationary phase which is immobilized on the support particles, or on the inner wall of the column tubing.
- The mobile phase is the phase which moves in a definite direction. It may be a liquid (LC and CEC), a gas (GC), or a supercritical fluid (supercritical-fluid chromatography, SFC). A better definition: The mobile phase consists of the sample being separated/ analyzed and the solvent that moves the sample through the column. In one case of HPLC the solvent consists of a carbonate/bicarbonate solution and the sample is the anions being separated. The mobile phase moves through the chromatography column (the stationary phase) where the sample interacts with the stationary phase and is separated.

- Preparative chromatography is used to purify sufficient quantities of a substance for further use, rather than analysis.
- The retention time is the characteristic time it takes for a particular analyte to pass through the system (from the column inlet to the detector) under set conditions.
- The sample is the matter analysed in chromatography. It may consist of a single component or it may be a mixture of components. When the sample is treated in the course of an analysis, the phase or the phases containing the analytes of interest is/are referred to as the sample whereas everything out of interest separated from the sample before or in the course of the analysis is referred to as waste.
- The solute refers to the sample components in partition chromatography.
- The solvent refers to any substance capable of solubilizing other substance, and especially the liquid mobile phase in LC.
- The stationary phase is the substance which is fixed in place for the chromatography procedure. Examples include the silica layer in thin layer chromatography.

TECHNIQUES BY CHROMATOGRAPHIC BED SHAPE

Column chromatography is a separation technique in which the stationary bed is within a tube. The particles of the solid stationary phase or the support coated with a liquid stationary phase may fill the whole inside volume of the tube (packed column) or be concentrated on or along the inside tube wall leaving an open, unrestricted path for the mobile phase in the middle part of the tube (open tubular column). Differences in rates of movement through the medium are calculated to different retention times of the sample.

In 1978, W. C. Still introduced a modified version of column chromatography called flash column chromatography (flash). The technique is very similar to the traditional column chromatography, except for that the solvent is driven through the column by applying positive pressure. This allowed most separations to be performed in less than 20 minutes, with improved separations compared to the old method.

Modern flash chromatography systems are sold as pre-packed plastic cartridges, and the solvent is pumped through the cartridge. Systems may also be linked with detectors and fraction collectors providing automation. The introduction of gradient pumps resulted in quicker separations and less solvent usage. In expanded bed adsorption, a fluidized bed is used, rather than a solid phase made by a packed bed. This allows omission of initial clearing steps such as centrifugation and filtration, for culture broths or slurries of broken cells.

PLANAR CHROMATOGRAPHY

Planar chromatography is a separation technique in which the stationary phase is present as or on a plane. The plane can be a paper, serving as such or impregnated by a substance as the stationary bed (paper chromatography) or a layer of solid particles spread on a support such as a glass plate (thin layer chromatography).

Paper Chromatography

Paper chromatography is a technique that involves placing a small dot of sample solution onto a strip of *chromatography paper*. The paper is placed in a jar containing a shallow layer of solvent and sealed. As the solvent rises through the paper it meets the sample mixture which starts to travel up the paper with the solvent. Different compounds in the sample mixture travel different distances according to how strongly they interact with the paper.

This paper is made of cellulose, a polar molecule, and the compounds within the mixture travel farther if they are non-polar. More polar substances bond with the cellulose paper more quickly, and therefore do not travel as far. This process allows the calculation of an R_f value and can be compared to standard compounds to aid in the identification of an unknown substance.

Thin Layer Chromatography

Thin layer chromatography (TLC) is a widely-employed laboratory technique and is similar to paper chromatography. However, instead of using a stationary phase of paper, it involves a stationary phase of a thin layer of adsorbent like silica gel, alumina, or cellulose on a flat, inert substrate. Compared to paper, it has the advantage of faster runs, better separations, and the choice between different adsorbents.

Different compounds in the sample mixture travel different distances according to how strongly they interact with the adsorbent. This allows the calculation of an R_f value and can be compared to standard compounds to aid in the identification of an unknown substance. For even better resolution and to allow for quantitation, high-performance TLC can be used.

TECHNIQUES BY PHYSICAL STATE OF MOBILE PHASE

Gas chromatography (GC), also sometimes known as Gas-Liquid chromatography, (GLC), is a separation technique in which the mobile phase is a gas. Gas chromatography is always carried out in a column, which is typically "packed" or "capillary". Gas chromatography (GC) is based on a partition equilibrium of analyte between a solid stationary phase (often a liquid silicone-based material) and a mobile gas (most often Helium). The stationary phase is adhered to the inside of a small-diameter glass tube (a capillary column) or a solid matrix inside a larger metal tube (a packed column). It is widely used in analytical chemistry; though the high temperatures used in GC make it unsuitable for high molecular weight biopolymers or proteins (heat will denature them), frequently encountered in biochemistry, it is well suited for use in the petrochemical, environmental monitoring, and industrial chemical fields. It is also used extensively in chemistry research.

Liquid Chromatography

Liquid chromatography (LC) is a separation technique in which the mobile phase is a liquid. Liquid chromatography can be carried out either in a column or a plane. Present day liquid chromatography that generally utilizes very small packing particles and a relatively high pressure is referred to as high performance liquid chromatography (HPLC).

In the HPLC technique, the sample is forced through a column that is packed with irregularly or spherically shaped particles or a porous monolithic layer (stationary phase) by a liquid (mobile phase) at high pressure. HPLC is historically divided into two different sub-classes based on the polarity of the mobile and stationary phases.

Technique in which the stationary phase is more polar than the mobile phase (e.g. toluene as the mobile phase, silica as the stationary phase) is called normal phase liquid chromatography (NPLC) and the opposite (e.g. water-methanol mixture as the mobile phase and C18 = octadecylsilyl as the stationary phase) is called reversed phase liquid chromatography (RPLC). Ironically the "normal phase" has fewer applications and RPLC is therefore used considerably more. Specific techniques which come under this broad heading are listed below. It should also be noted that the following techniques can also be considered fast protein liquid chromatography if no pressure is used to drive the mobile phase through the stationary phase.

Affinity Chromatography

Affinity chromatography is based on selective non-covalent interaction between an analyte and specific molecules. It is very specific, but not very robust. It is often used in biochemistry in the purification of proteins bound to tags. These fusion proteins are labelled with compounds such as His-tags, biotin or antigens, which bind to the stationary phase specifically. After purification, some of these tags are usually removed and the pure protein is obtained.

SUPERCRITICAL FLUID CHROMATOGRAPHY

Supercritical fluid chromatography is a separation technique in which the mobile phase is a fluid above and relatively close to its critical temperature and pressure.

TECHNIQUES BY SEPARATION MECHANISM

ION EXCHANGE CHROMATOGRAPHY

Ion exchange chromatography utilizes ion exchange mechanism to separate analytes. It is usually performed in columns but the mechanism can be benefited also in planar mode. Ion exchange chromatography uses a charged stationary phase to separate charged compounds including amino acids, peptides, and proteins. In conventional methods the stationary phase is an ion exchange resin that carries charged functional groups which interact with oppositely charged groups of the compound to be retained. Ion exchange chromatography is commonly used to purify proteins using FPLC.

SIZE EXCLUSION CHROMATOGRAPHY

Size exclusion chromatography (SEC) is also known as gel permeation chromatography (GPC) or gel filtration chromatography and separates molecules according to their size (or more accurately according to their hydrodynamic diameter or hydrodynamic volume). Smaller molecules are able to enter the pores of the media and, therefore, take longer to elute, whereas larger molecules are excluded from the pores and elute faster. It is generally a low resolution chromatography technique and thus it is often reserved for the final, "polishing" step of a purification. It is also useful for determining the tertiary structure and quaternary structure of purified proteins, especially since it can be carried out under native solution conditions.

SPECIAL TECHNIQUES

Reversed-phase Chromatography

Reversed-phase chromatography is an elution procedure used in liquid chromatography in which the mobile phase is significantly more polar than the stationary phase.

Two-dimensional Chromatography

In some cases, the chemistry within a given column can

be insufficient to separate some analytes. It is possible to direct a series of unresolved peaks onto a second column with different physico-chemical (Chemical classification) properties. Since the mechanism of retention on this new solid support is different from the first dimensional separation, it can be possible to separate compounds that are indistinguishable by one-dimensional chromatography.

Simulated Moving-Bed Chromatography

Fast protein liquid chromatography (FPLC) is a term applied to several chromatography techniques which are used to purify proteins. Many of these techniques are identical to those carried out under high performance liquid chromatography, however use of FPLC techniques are typically for preparing large scale batches of a purified product.

Countercurrent Chromatography

Countercurrent chromatography (CCC) is a type of liquid-liquid chromatography, where both the stationary and mobile phases are liquids. It involves mixing a solution of liquids, allowing them to settle into layers and then separating the layers.

Chiral Chromatography

Chiral chromatography involves the separation of stereoisomers. In the case of enantiomers, these have no chemical or physical differences apart from being three dimensional mirror images.

Conventional chromatography or other separation processes are incapable of separating them. To enable chiral separations to take place, either the mobile phase or the stationary phase must themselves be made chiral, giving differing affinities between the analytes. Chiral chromatography HPLC columns (with a chiral stationary phase) in both normal and reversed phase are commercially available.

Index

R

S

T

U

V

W

Y